Simulation and Verification of Electronic and Biological Systems

Peng Li • Luís Miguel Silveira • Peter Feldmann
Editors

Simulation and Verification of Electronic and Biological Systems

Springer

Editors
Peng Li
Dept. of Electrical and Comp. Eng.
Texas A&M University
3259 TAMU
77843 College Station
USA
pli@tamu.edu

Peter Feldmann
IBM T.J. Watson Research Center
Kitchawan Rd. 1101
10598 Yorktown Heights
USA
feldmann@us.ibn.com

Luís Miguel Silveira
Dept. of Electrical and Comp. Eng.
INESC ID/IST - TU Lisbon
Rua Alves Redol 9
1000-029 Lisboa
Portugal
lms@inesc-id.pt

ISBN 978-94-007-0148-9 e-ISBN 978-94-007-0149-6
DOI 10.1007/978-94-007-0149-6
Springer Dordrecht Heidelberg London New York

Cover design: eStudio Calamar S.L.

Printed on acid-free paper

Springer is part of Springer Science+Business Media (www.springer.com)

Foreword

The first fifty years of the semiconductor industry have seen a remarkable sustained exponential rate of growth that is unprecedented! The industry continues to make ever smaller devices, more complex fabrication processes, and larger and more capable systems, all the while continuing to reduce the cost per component and per computation—a truly impressive achievement for the scientists and engineers who work in this area.

Creating integrated circuits is different from other fields of engineering in one striking aspect: it does not rely on prototyping! An automotive engineer will, as a matter of course, build a working prototype of his car, measure its performance in a wind tunnel, drive it around a racetrack, and even crash test it. A civil engineer will build a scale model of a bridge and study its behavior under various loading or environmental conditions. But an integrated circuit design engineer does not have that luxury—he must get the design "right" the first time, and he does it by relying almost exclusively on computer simulation of the integrated circuit.

This near-exclusive reliance on simulation is somewhat unique to the semiconductor industry, though its economic benefits—demonstrated by our phenomenal success—are now coaxing other engineering disciplines to move in the same direction. At its core, the prediction of circuit performance via simulation requires three things:

1. Abstractions of the distributed (albeit tiny) devices that allow them to be viewed purely via their interface terminals. We refer to these abstractions as "compact models".
2. Conservation laws for charge and energy that mathematically describe how a circuit comprised of a collection of devices behaves. We refer to these as Kirchhoffs current and voltage law.
3. Numerical algorithms that allow the efficient simulation of the behavior of systems comprised of compact models and obeying conservation laws.

We take these concepts for granted, but it is useful to remind ourselves occasionally of how difficult this task is in general, and of how we are standing on the shoulders of the giants in our field who blazed the trail we now so confidently tread.

Because of the huge investment that the semiconductor industry has put into R&D in the area of modeling and simulation, the state of our art in this area is unrivaled in other engineering and scientific disciplines. Our deep level of understanding in this area, as well as our demonstrated ability to make accurate predictions of systems comprising billions of interacting components is the envy of other fields, so much so that the basic knowledge and algorithms developed to allow the simulation of semiconductors are now "leaking" to other fields (as illustrated in parts of this work exploring biological systems). This book demonstrates the high level of achievement we have reached, and shines the light on a future in which this community will continue to creatively find solutions in the semiconductor area and beyond.

Austin, Texas, USA, October 2010 *Sani R. Nassif*

Preface

Simulation is a fundamental capability for understanding, analyzing and designing an extremely wide range of engineering and biological circuits and systems. As an example, electronic circuit simulation techniques and tools such as SPICE have been a fundamental workhorse to enable the development of integrated circuits in the past several decades. As design complexity drastically rises and manufacturing technologies evolve into the nanometer regime, there is an unsaturated demand for delivering higher accuracy, faster speed and novel simulation capabilities. As such, the notion of simulation should also be interpreted broadly to include complementary modeling and verification techniques that serve the same goal of facilitating complex system design. On a broader horizon, interesting new simulation challenges emerge in biological applications, whose solutions may be assisted by exploiting analysis and tools in electrical engineering and contribute to the development of simulation techniques at large.

With this theme, this book showcases selected works presented at the Circuit and Multi-Domain Simulation Workshop, co-located with the IEEE/ACM International Conference on Computer-Aided Design (ICCAD) in San Jose, California, USA, on November 5, 2009. The workshop drew a large audience from academia, industry and national research laboratories and forged stimulating discussions in the community. The nine chapters of this book provide a comprehensive discussion of recent developments in simulation, modeling and verification of integrated circuits and biological systems. Specific topics include large scale parallel circuit simulation, industrial practice of fast SPICE simulation, structure-preserving model order reduction of interconnects, advanced simulation techniques for oscillator networks, dynamic stability of static memories and biological systems as well as verification of analog integrated circuits.

We sincerely thank the speakers and attendees of the workshop, who had made the meeting a great success. The effort and patience of the chapter contributors are gratefully acknowledged. Without it, it would not be possible for us to have such a wonderful book. We also want to thank Synopsys Inc. and the ACM SIGDA Physical Design Technical Committee for providing financial sponsorship for the workshop, and the ICCAD organizing committee for their cooperation.

Finally, the assistance received from Springer, particularly, that from Mark de Jongh and Cindy Zitter, during the publication of this book is acknowledged.

College Station, Lisbon and Yorktown Heights
October 2010

Peng Li
L. Miguel Silveira
Peter Feldmann

Contents

List of Contributors

Joshua Apgar
Systems Biology Group, Boehringer Ingelheim Pharmaceuticals, Ridgefield, CT 06877, USA, e-mail: apgar@mit.edu

Prateek Bhansali
Department of Electrical Engineering and Computer Sciences, University of California, Berkeley, CA 94720-1770, USA, e-mail: bhansali@berkeley.edu

Anya Castillo
Technology Policy Program, Massachusetts Institute of Technology, Cambridge, MA 02139, USA, e-mail: anyac@mit.edu

Henry Chang
Designer's Guide Consulting, 101 First Street #150, Los Altos, CA 94022, USA, e-mail: henry@designers-guide.com

Wei Dong
Texas Instruments, 12500 TI Boulevard, MS 8714, Dallas, TX 75243, USA, e-mail: weidong@ti.com

Roland W. Freund
Department of Mathematics, University of California at Davis, One Shields Avenue, Davis, CA 95616, USA, e-mail: freund@math.ucdavis.edu

Lars Hedrich
Electronic Design Methodology, Institute for Computer Science, Goethe University of Frankfurt/Main, Robert-Mayer-Str. 11-15, D-60325 Frankfurt/Main, Germany, e-mail: hedrich@em.cs.uni-frankfurt.de

Garng M. Huang
Department of Electrical and Computer Engineering, Texas A&M University, College Station, TX 77843, USA, e-mail: ghuang@neo.tamu.edu

Eric R. Keiter
Electrical and Microsystems Modeling Department, Sandia National Laboratories, P.O. Box 5800, Albuquerque, NM 87185-0316, USA, e-mail: erkeite@sandia.gov

Ken Kundert
Designer's Guide Consulting, 101 First Street #150, Los Altos, CA 94022, USA, e-mail: ken@designers-guide.com

Peng Li
Department of Electrical and Computer Engineering, Texas A&M University, College Station, TX 77843, USA, e-mail: pli@tamu.edu

Elebeoba E. May
Sandia National Laboratories, P.O. Box 5800, Albuquerque, NM 87185-0316, USA, e-mail: eemay@sandia.gov

Eric L. Rankin
Electrical and Microsystems Modeling Department, Sandia National Laboratories, P.O. Box 5800, Albuquerque, NM 87185-0316, USA, e-mail: elranki@sandia.gov

Michał Rewieński
Synopsys Inc., 700 E. Middlefield Rd., Mountain View, CA 94043, USA, e-mail: michalr@synopsys.com

Jaijeet Roychowdhury
Department of Electrical Engineering and Computer Sciences, University of California, Berkeley, CA 94720-1770, USA, e-mail: jr@berkeley.edu

Thomas V. Russo
Electrical and Microsystems Modeling Department, Sandia National Laboratories, P.O. Box 5800, Albuquerque, NM 87185-0316, USA, e-mail: tvrusso@sandia.gov

Richard L. Schiek
Electrical and Microsystems Modeling Department, Sandia National Laboratories, P.O. Box 5800, Albuquerque, NM 87185-0316, USA, e-mail: rlschiek@sandia.gov

Sebastian Steinhorst
Electronic Design Methodology, Institute for Computer Science, Goethe University of Frankfurt/Main, Robert-Mayer-Str. 11-15, D-60325 Frankfurt/Main, Germany, e-mail: steinhorst@em.cs.uni-frankfurt.de

Heidi K. Thornquist
Electrical and Microsystems Modeling Department, Sandia National Laboratories, P.O. Box 5800, Albuquerque, NM 87185-0316, USA, e-mail: hkthorn@sandia.gov

Bruce Tidor
Department of Electrical Engineering and Computer Science and the Department of Biological Engineering, Massachusetts Institute of Technology, Cambridge, MA 02139, USA, e-mail: tidor@mit.edu

Jared Toettcher
Department of Cellular and Molecular Pharmacology, University of California, San Francisco, CA 94158, USA, e-mail: jared.toettcher@ucsf.edu

Jacob White
Department of Electrical Engineering and Computer Science and the Department of Biological Engineering, Massachusetts Institute of Technology, Cambridge, MA 02139, USA, e-mail: white@mit.edu

Parallel Transistor-Level Circuit Simulation

Eric R. Keiter, Heidi K. Thornquist, Robert J. Hoekstra, Thomas V. Russo, Richard L. Schiek and Eric L. Rankin

Abstract With the advent of multi-core technology, inexpensive large-scale parallel platforms are now widely available. While this presents new opportunities for the EDA community, traditional transistor-level, SPICE-style circuit simulation has unique parallel simulation challenges. Here the Xyce Parallel Circuit Simulator is described, which has been designed from the "from-the-ground-up" to be distributed memory-parallel. Xyce has demonstrated scalable circuit simulation on hundreds of processors, but doing so required a comprehensive parallel strategy. This included the development of new solver technologies, including novel preconditioned iterative solvers, as well as attention to other aspects of the simulation such as parallel file I/O, and efficient load balancing of device evaluations and linear systems. Xyce relies primarily upon a message-passing (MPI-based) implementation, but optimal scalability on multi-core platforms can require a combination of message-passing and threading. To accommodate future parallel platforms, software abstractions allowing adaptation to other parallel paradigms are part of the Xyce design.

Eric R. Keiter, Heidi K. Thornquist, Thomas V. Russo, Richard L. Schiek and Eric L. Rankin
Electrical and Microsystems Modeling Department
Sandia National Laboratories
P.O. Box 5800, Albuquerque, NM 87185-0316, USA
e-mail: `erkeite@sandia.gov`, `hkthorn@sandia.gov`, `tvrusso@sandia.gov`, `rlschiek@sandia.gov`, `elranki@sandia.gov`

Robert J. Hoekstra
Applied Mathematics & Applications Department
Sandia National Laboratories
P.O. Box 5800, Albuquerque, NM 87185-0318, USA
e-mail: `rjhoeks@sandia.gov`

Sandia National Laboratories is a multi-program laboratory managed and operated by Sandia Corporation, a wholly owned subsidiary of Lockheed Martin Corporation, for the U.S. Department of Energy's National Nuclear Security Administration under contract DE-AC04-94AL85000.

P. Li et al. (eds.), *Simulation and Verification of Electronic and Biological Systems*,
DOI 10.1007/978-94-007-0149-6_1, © Springer Science+Business Media B.V. 2011

1 Introduction

At modern technology nodes, analog style SPICE-accurate simulation can be a significant (and prohibitive) development bottleneck. Traditional circuit simulation, originally made popular by the Berkeley SPICE program [26], does not scale well beyond tens of thousands of unknowns, due to the use of direct matrix solver methods.

A number of algorithms for Fast-SPICE tools have been developed which allow for faster, larger-scale circuit simulation. Often based on circuit-level partitioning algorithms [1, 3, 27, 34], such tools can be applied to much larger problems, but the approximations inherent to such algorithms can break down under some circumstances. In particular, for state-of-the-art modern VLSI design, high levels of integration between functional modules and interconnects are subject to prohibitive parasitic effects, and can render such tools unreliable.

Recent development of inexpensive computer clusters, as well as multi-core technology, has resulted in significant interest for efficient parallel circuit simulation. Parallel "true-SPICE" circuit simulation has been investigated previously, including Frölich [13], who relied on a multi-level Newton approach in the Titan simulator; Basermann [8], who used a Schur-complement based preconditioner; and Peng et al. [28] used a domain decomposition approach and relied on a combination of direct and iterative solvers. Recently, interest has developed around parallel SPICE acceleration using graphical processing units (GPUs) [15].

Parallel circuit simulation requires integration of large and small scale parallelism throughout the entire circuit simulation flow. In this paper, parallel algorithms for circuit simulation, and their implementation in a production simulator, Xyce, are discussed. Xyce [4] is a simulator designed "from-the-ground-up" to be distributed memory-parallel, and is targeted at a spectrum of parallel platforms, from high-end supercomputers, to large clusters, to multi-core desktops. It relies primarily upon a message-passing implementation (MPI) [14], but optimal scalability on multi-core technology can require combined message-passing and threading. Xyce uses software abstractions that allow the simulator to adapt to other parallel paradigms.

2 Background

Circuit simulation adheres to a general flow, as shown in Fig. 1. The circuit, described in a netlist file, is transformed via modified nodal analysis (MNA) into a set of nonlinear differential algebraic equations (DAEs)

$$\frac{dq(x(t))}{dt} + f(x(t)) = b(t), \tag{1}$$

where $x(t) \in \mathbb{R}^N$ is the vector of circuit unknowns, q and f are functions representing the dynamic and static circuit elements (respectively), and $b(t) \in \mathbb{R}^M$ is the input

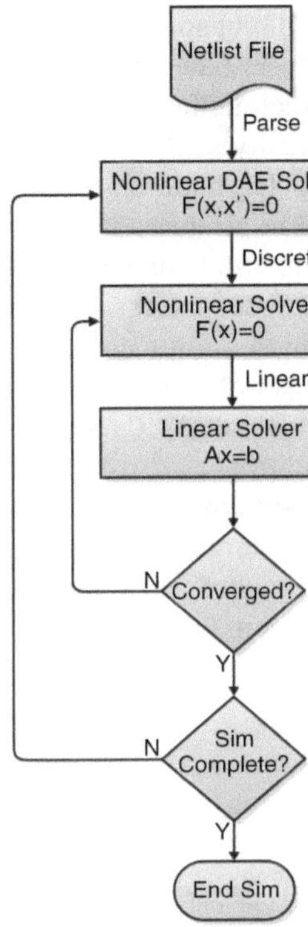

Fig. 1 General circuit simulation flow

vector. For any analysis type, the initial starting point is this set of DAEs. The numerical approach employed to compute solutions to (1) is predicated by the analysis type.

Transient and DC analysis are two commonly used simulation modes, in which the set of equations (1) are solved by numerical integration methods corresponding to the nested solver loop in Fig. 1. For transient analysis, linear systems are of the form:

$$(G+Q/\delta t)\delta x = (b-f)/\delta t \qquad (2)$$

involving the conductance matrix $G(t) = \frac{df}{dx}(x(t))$, and the capacitance matrix $Q(t) = \frac{dq}{dx}(x(t))$. For DC analysis, the q terms are not present, so (1) is reduced to the nonlinear equation $f(x) = 0$, and the linear system is simplified to:

$$G\,\delta x = -f(x). \qquad (3)$$

For transient and DC analysis, the computational expense is in repeatedly solving linear systems of equations given by (2) or (3), respectively. These linear systems are typically sparse, have heterogeneous non-symmetric structure, and are often ill-conditioned. As such, iterative matrix solvers have historically not been the first choice for circuit simulation, and direct sparse solvers [10, 21] have been the industry standard approach. Direct solvers have the advantage of reliability and ease of use, and for smaller problems direct methods are usually faster than iterative methods. However, direct solvers typically scale poorly with problem size and become impractical when the linear system has hundreds of thousands of unknowns or more.

Despite the problems inherent to circuit matrices, iterative solvers have the potential to be a scalable solution method for large-scale linear systems with lower algorithmic complexity. They are not as easy to use as direct solvers because their effectiveness is dependent upon finding an adequate preconditioner. However, progress has been made on the use of iterative methods in transient circuit analysis; notably Basermann [8] and Bomhof [9], both of whom relied on distributed Schur-complement based preconditioners. Also, multi-grid methods have successfully been applied to power-grid simulation [32]. Recently, a new preconditioning strategy has been developed to generate effective preconditioners for performing transient analysis using matrix structure found during DC analysis [16]. Nonetheless, for conventional transient circuit simulation, iterative matrix solvers have yet to be widely used in production tools.

3 Parallelism Opportunities in Circuit Simulation

Parallelism can be integrated into every step of the circuit simulation flow shown in Fig. 1. Furthermore, at every step, parallelism can be achieved through both coarse-scale (multi-processor) and fine-scale (multi-threaded) approaches. A composition of these two approaches will provide circuit simulation with the best performance impact on the widest variety of parallel platforms.

This section contains a discussion of the parallelism opportunities exploited by the Xyce simulation flow. Section 3.1 will focus on the parallel netlist parser, while Sect. 3.2 describes parallelism in the nonlinear and linear solvers. The majority of the computational time is spent in device evaluations and linear solvers, so Sect. 3.2 will focus on parallelism pertaining to those specific tasks.

3.1 Parallel Netlist Parser

An efficient parallel netlist parser is essential for the simulation of very large circuits. The parser is the gateway through which the devices and network topology are established and, if inefficient in computational time or memory, can easily become the Achilles' heel of a circuit simulator. Developing a netlist parser to work on a wide variety of platforms, from desktops to supercomputers, can be a harrowing task because of parallel file I/O. As such, the netlist parser in Xyce has had to address two file system scenarios: inhomogeneous and distributed, homogeneous file systems.

3.1.1 Inhomogeneous File System

An inhomogeneous file system, often found in networked computer clusters, is the most general type of parallel file system. For such a file system, it cannot be assumed that each node has access to the same directories and files, so the netlist has to be streamed on one processor. This presents an immediate parallel bottleneck, both in terms of performance as well as memory. The approach taken in Xyce for this sort of file system involves minimally processing the netlist prior to distributing devices and analysis information to the other processors.

Parsing is accomplished in multiple passes of the netlist. In the first pass, the netlist is dynamically flattened and a total device instance count is obtained. This allows an initial, naive, parallel partitioning of devices to be determined based on D/N, where D = number of devices and N = number of nodes. On the subsequent pass, raw netlist data for each device is broadcast sequentially to the remaining nodes or processors. After this process is completed, the devices are instantiated and initialized locally on each node. This approach does not completely remove the parallel bottleneck inherent to streaming the netlist file on one processor. However, it enables the parsing process to be mostly scalable in terms of memory usage.

The initial D/N partition determines the load balance for device evaluation. For large problems, a naive partition (where each device is given an identical weight) is often adequate, as the runtime is dominated by the linear solve at each step. However, a more optimal device evaluation partition can easily be created by using weightings.

3.1.2 Distributed Homogeneous File System

A homogeneous parallel file system is a more common file system on modern parallel architectures. Though less general, it is in principle more scalable. Similar to the heterogeneous case, an initial pass through the netlist is necessary, to determine device count, and also to set file pointers for each node. As each node has full access to the file system, parts of the netlist can be streamed in to each node directly, rather than streamed in on one processor and then communicated. While this has the potential to be scalable in terms of operations as well as memory, the bookkeeping can be prohibitive for very hierarchical netlists. As with the heterogeneous file system, the initial device evaluation partition is determined as part of the parallel netlist parsing process.

3.2 Parallel Approach for Nested Solver Loop

In Xyce, the parallel approach taken for the nested solver loop is designed to account for circuit heterogeneity, so optimal parallel load balance for device evaluation (matrix and residual vector assembly) will likely differ from that of the linear solution

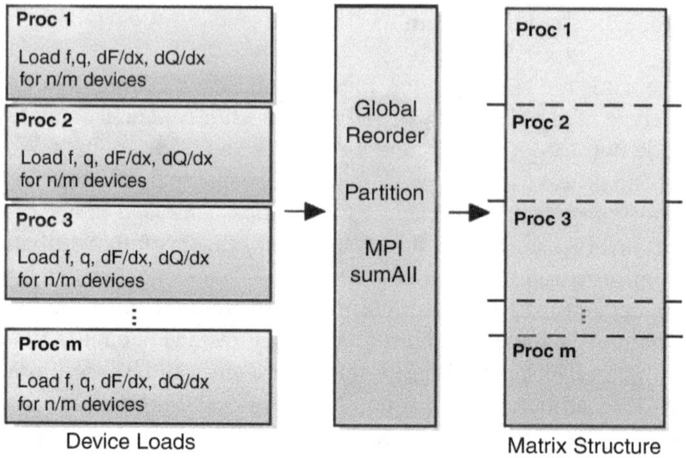

Fig. 2 Different load balance/partitioning for device evaluation and linear solve

phase. Device evaluation and linear solves each happen once per Newton iteration, so over the course of a long run the combined cost of both will comprise the bulk of the wall clock simulation time.

The relative amount of time spent in each phase is problem-dependent. For smaller problems, the device evaluation phase should dominate run time. As the problem size increases, the linear solve phase will dominate, as it should scale super-linearly, while the device evaluations should scale linearly. This is because linear solution methods (whether they be direct or iterative) are generally communication intensive, while the communication volume required during the device evaluations is relatively small.

As a result, the device evaluation phase can be balanced by taking into account only the computational work required, while the matrix partitioning has to minimize communication volume. How this communication volume is measured, and how it is optimized is an active area of research for many types of numerical simulation problems. Since the device evaluation and solve phases have different load balance requirements, Xyce has been designed to have completely different parallel partitioning for each. A simplified representation of this is shown in Fig. 2.

3.3 Device Evaluation

In Fig. 2, the left column represents the device evaluation procedure. Communication costs for this are relatively low and insensitive to load balance. At the beginning of each device evaluation, solution and state vector values are needed by each device, and off-processor values are communicated as necessary. At the end of the device evaluation, before final assembly, residual and Jacobian matrix values must be communicated to fill the final linear system. Thus, the communication cost is relatively cheap, and the main metric considered is the computational cost. For many circuits of interest, a naive load balance, in which the total number of devices is evenly divided among the available processors will demonstrate very good parallel scaling. For circuits that are very heterogeneous, weights can be applied to different device types to achieve a better balance.

The middle box in Fig. 2 represents the communication necessary to accommodate both load balances. This is dependent upon the partitioning of the linear system, which is a much more difficult and complex issue. Unlike the device evaluation, a naive partitioning will generally not suffice.

3.4 Linear Solvers for Circuit Simulation

Solving the linear systems of equations, (2) and (3), generated through circuit simulation is often challenging. These linear systems are sparse, typically have heterogeneous non-symmetric structure, and are often ill-conditioned. Direct sparse linear solvers provide a reliable, easy-to-use solution method that is preferred, industry-wide, over iterative solvers. For smaller linear systems, this is understandable, because direct solvers are usually faster than their iterative counterparts. However, when the linear system has hundreds of thousands of unknowns or more, direct solvers become less practical as they suffer from poor scaling. So any advantage gained through a scalable device evaluation procedure, is lost through the linear solver. This situation only becomes more pronounced as the problem, or circuit, size increases.

Iterative solvers, like GMRES [30], are scalable and robust for other types of physical problems, but not generally for circuit simulation. This issue results from the inability to generate effective, general-purpose, scalable preconditioners. The heterogeneity in circuit matrices presents an exceptional challenge for creating a preconditioner that approximates the coefficient matrix well, to accelerate convergence, and is inexpensive to apply and scalable. Xyce takes advantage of several common matrix properties when constructing a preconditioner for (2) or (3). In the rest of this section the algorithms that take advantage of these properties will be presented, including: singleton removal, block-triangular permutation, and parallel partitioning. This will be followed by a discussion of the iterative linear solver strategies that are used in Xyce.

3.4.1 Singleton Removal

Conductance and capacitance matrices are sparse, but typically contain dense rows and columns. Such structural matrix features are problematic for parallelism because they have the potential to increase communication costs dramatically. A crucial observation [8] is that the dense rows (or columns) correspond to columns (or

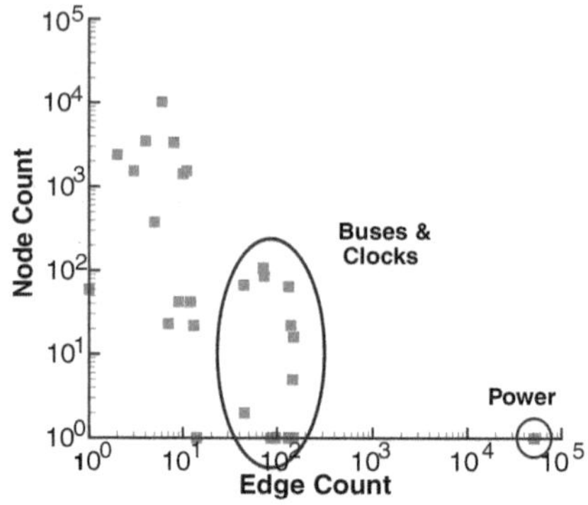

Fig. 3 Example IC connectivity histogram

rows) with one and only one non-zero entry, called a singleton. Eliminating singleton rows and columns, a standard practice in the original sparse direct methods for circuits (Sect. 6.3.1 of [21]), also eliminates the dense rows and columns. These dense rows and columns typically result from power supply and ground nodes, which are common to digital circuits. Other features such as clock nodes, while not as highly connected as power nodes, still have a high enough connectivity to cause problems. A histogram illustrating the connectivity for a sample integrated circuit (IC) is depicted in Fig. 3.

Dense rows and columns can easily be removed from circuit matrices as pre- and post-solve steps because the corresponding columns or rows will only have one non-zero entry [8]. The procedure, referred to as "singleton removal", is used by Xyce as a first step for solving (2) and (3). The pre-processing step for handling row singletons by removing the independent variable (x_j), is given on the left in Fig. 4. While the post-processing step for handling column singletons, where the variable (x_j) is fully dependent, is given on the right in Fig. 4.

Singleton removal is a successful strategy, but only when the nodes are connected to ideal sources, and thus can essentially be removed from the system of equations by inspection, as their values are known. In practice, highly connected nodes may be non-ideal due to parasitic elements, and under this circumstance, the singleton removal algorithm can fail to detect them. A mitigation strategy designed to preserve singleton removal, based on multi-level Newton methods is presented in Sect. 4.

3.4.2 BTF Reordering

The conductance matrix G used during DC analysis can be reducible [10], permutable to a block triangular matrix with small diagonal blocks. Exploiting this block triangular form (BTF) often gives great performance gains both for direct and iterative solvers. Direct solvers only factor the tiny diagonal blocks, handling the off-diagonal blocks in the substitution phase. However, the challenge is in performing transient analysis, when the linear systems are not reducible. Iterative solvers

$$\begin{bmatrix} & a_{1j} & \\ & \vdots & \\ 0 \cdots & a_{ij} & \cdots 0 \\ & \vdots & \\ & a_{nj} & \end{bmatrix} \begin{bmatrix} x_1 \\ \vdots \\ x_j \\ \vdots \\ \vdots \\ x_n \end{bmatrix} = \begin{bmatrix} b_1 \\ \vdots \\ \vdots \\ b_i \\ \vdots \\ b_n \end{bmatrix} \Rightarrow x_j = b_i / a_{ij}$$

$$\begin{bmatrix} & 0 & \\ & \vdots & \\ a_{i1} \cdots & a_{ij} & \cdots a_{in} \\ & \vdots & \\ & 0 & \end{bmatrix} \begin{bmatrix} x_1 \\ \vdots \\ x_j \\ \vdots \\ \vdots \\ x_n \end{bmatrix} = \begin{bmatrix} b_1 \\ \vdots \\ \vdots \\ b_i \\ \vdots \\ b_n \end{bmatrix} \Rightarrow x_j = \left(b_i - \sum_{k \neq j} a_{ik} x_k \right) / a_{ij}$$

Fig. 4 Singleton removal for rows (left) and columns (right)

can take advantage of this block structure to generate preconditioners that are effective for both DC and transient analysis [16].

The permutation to block triangular form, via the Dulmage Mendelsohn decomposition, has two steps. First, a maximum matching permutation generates a matrix with a zero-free diagonal. Second, a topological sort finds the strongly-connected components of the associated directed graph. In Xyce, computing the permutation to block triangular form is a serial bottleneck because there is no parallel software available. However, both steps required to compute this transformation can probably be parallelized. The first step requires a bipartite matching, which is the most difficult to parallelize [24], but most device models produce Jacobian matrices with very few zeros on the diagonal. The second step finds strongly connected components and recent work shows this can be done in parallel [25]. For the time being, computing the BTF permutation in serial is acceptable because the same reordering can be used for each Newton step of the DC and transient analysis. Thus, the cost is amortized over the entire simulation.

3.4.3 Parallel Partitioning

An effective sparse matrix partitioning, where the goal is to reduce the communication in sparse matrix-vector multiplication, is essential to iterative matrix solvers. Furthermore, it has been demonstrated for a variety of physical simulation problems [11, 33] that a good parallel partition can enhance the performance of preconditioned iterative matrix solvers. Graph-based partitioning algorithms have been popular for some time [20] and became computationally feasible for large problems with the advent of multi-level methods [17]. In recent years, hypergraph partitioning has shown a lot of promise [12], in part because it relies upon more accurate metrics for measuring parallel communication costs. This will give an exact estimate of the costs required for a matrix-vector multiply, which is the main communication expense for iterative methods like GMRES. Interestingly enough, hypergraph-based partitioning has also been demonstrated to be an effective algorithm for optimizing circuit layout for path-based delay minimization [18].

Sparse matrix partitioning is performed in Xyce using either a graph or hypergraph partitioner via ParMETIS [19] or Zoltan [12], respectively. Either approach generates a 1D partition, which partitions only the rows, assigning each row to only one processor. Partitioning is computationally expensive, but the graph for conductance and capacitance matrices is static over the course of the simulation. Thus, the partitioning can be reused, amortizing the cost over the entire simulation.

3.4.4 Iterative Linear Solver Strategies

In Xyce, the previously discussed matrix transformations are combined with algebraic preconditioning techniques to generate an iterative linear solver strategy. The two strategies that have proven to be effective in preconditioning circuit matrices will be presented in this section. In both, the linear solver is a block diagonal preconditioned GMRES [30] method, in which the number of "blocks" is the number of cores or processors used. To avoid confusion these "blocks" are referred to as subdomains. A block diagonal preconditioner applies the exact (Jacobi) or approximate inverse of the subdomain matrix as its preconditioner. This preconditioner requires no communication to perform the factorization and in its application, which makes it suitable for parallel computation.

The first strategy is a general-purpose domain decomposition (DD) strategy, and is illustrated in Fig. 5. Domain decomposition uses singleton removal, graph partitioning (ParMETIS [19]) on the resulting symmetrized graph to reduce communication, and then computes a local fill-reducing ordering (AMD [6]) on the block diagonal before performing a complete or incomplete LU (ILU) factorization. This is used as a preconditioner for GMRES [30].

The second linear solver strategy was recently developed and is based on the block triangular structure that is often found in conductance matrices [16]. When the BTF structure is present, this strategy generates an effective preconditioner for DC analysis that can also be used in transient analysis. The BTF linear solver strategy, as illustrated in Fig. 5, uses singleton removal and then permutes the resulting graph to block triangular form. The strongly connected components are then partitioned, and each subdomain is solved directly using KLU [31] as a preconditioner for GMRES [30].

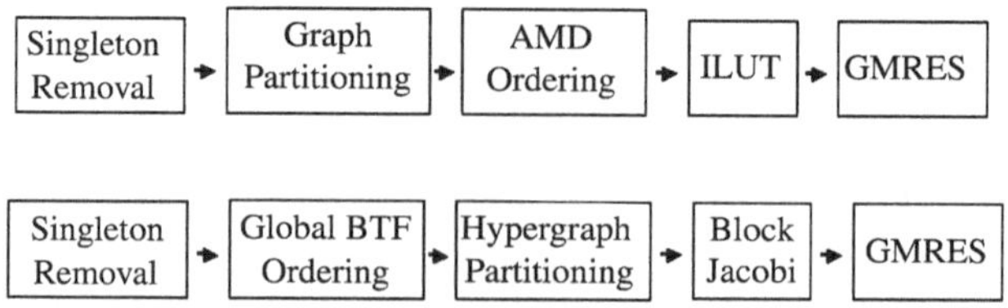

Fig. 5 Domain decomposition (top) and BTF (bottom) linear solver strategies

4 Graph Mitigation using Multi-level Newton Methods

Several of the linear solver steps described in Sect. 3.4 depend upon the circuit or Jacobian matrix having specific structure. For example, the singleton removal algorithm (Sect. 3.4.1) assumes that power and clock node voltages are set with ideal sources, with no intervening parasitic elements. Also, the BTF reordering (Sect. 3.4.2) only works if the circuit or matrix is sufficiently unidirectional, which is not the case for all circuits.

Often, it is possible to mitigate structural problems by strategically partitioning the nonlinear solver, and applying the multi-level Newton method [29]. In general, such methods have been used to enable a simulator to optimally apply ideal solver methods to different phases of a simulation. Peng et al. [28] used a multi-level Newton approach to enhance post-layout transistor-level circuit simulation. In this work, the problem was partitioned into linear and nonlinear partitions. The linear partition (due to post-layout parasitics), produced an SPD matrix that could be efficiently solved using the Conjugate-Gradient (CG) method, while the nonlinear partition produced a much smaller non-SPD matrix that was solved using direct methods.

For parallel circuit simulation, the multi-level Newton method can be applied to break up the graph structure of the circuit, with the goal of preserving the effectiveness of matrix ordering and other pre-processing steps. The method will be described next, followed by its application to non-ideal power supplies (Sect. 4.2).

4.1 Multi-level Newton Method

The multi-level Newton method has been described by numerous authors [29], and involves partitioning a problem into a tree structure, with a top level and subordinate lower levels. Circuits can often be very naturally partitioned based on subcircuits, but such a partition will not always fulfill the numerical objective which justifies using the method.

An illustration of a multi-level Newton partitioning, consisting of two levels, is shown in Fig. 6. The variables x_u, x_l and x_i represent solution variables for the upper level, lower level, and interface respectively.

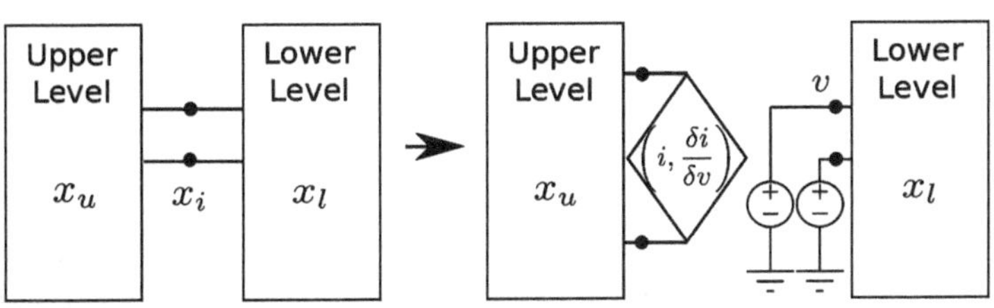

Fig. 6 Simple multi-level Newton decomposition, where x_u, x_l and x_i are the solution variables for the upper level, lower level, and interface, respectively

The multi-level method treats each level separately, with the lower level undergoing a full Newton solve for each iteration of the upper level. For circuit simulation, using the MNA formulation, most of the solution variables x will be nodal voltages. The interface nodal values x_i will be imposed into each lower level as fixed voltage boundary conditions, or independent voltage sources, as indicated in Fig. 6.

From the point of view of the upper level, the subordinate portions of the circuit, or lower levels, are replaced by macromodels, which provide Ohmic relationships similar to that of conventional compact device models. As such, each lower level subcircuit must provide currents and conductances to fill out their respective matrix stamps in the upper level system.

After each lower level solution has been obtained, that solution is then used to help construct the upper level linear system. Each macromodel provides a matrix stamp that is equal in size to the number of terminal connections, and each entry in that stamp is a conductance term, obtained by solving the following equation:

$$J_{mn} = \frac{\partial f_m}{\partial x_i} = \frac{\partial \hat{f}_m}{\partial x_i} + \frac{\partial f_m}{\partial x_l}\frac{\partial x_l}{\partial x_i} \tag{4}$$

where m is the row index and n is the column index. From the perspective of the top level circuit Jacobian, m corresponds to the KCL equation associated with the circuit node attached to the mth electrode of the lower problem, and n corresponds to the voltage variable for the circuit node attached to the nth electrode. The first term $\hat{f}_m/\partial x_i$ in (4) corresponds to the inner problem contribution to the circuit node equation on the matrix diagonal, and is zero if $m \neq n$.

The second term in (4) is the product of two vectors, $\partial f_m/\partial x_l$ and $\partial x_l/\partial x_i$. The vector $\partial f_m/\partial x_l$ of the second term of (4) corresponds to a sparse row of the lower problem Jacobian, or the derivatives of the terminal KCL equations with respect to the internal variables of the lower problem. The vector $\partial x_l/\partial x_i$ is determined from solving the linear system:

$$J_l \frac{\partial x_l}{\partial x_i} = -\frac{\partial f_l}{\partial x_i} \tag{5}$$

where x_i are interface variables, corresponding to nodal voltages at the terminals to the lower problem, J_l is the Jacobian of the lower level problem, and x_l are lower problem variables. The right hand side of (5), $\partial f_l/\partial x_i$ is a sparse vector that corresponds to a column of the full system Jacobian for the derivatives of lower system variables with respect to interface variables.

4.2 Preserving Singleton Removal

As mentioned in Sect. 3.4.1, singleton removal can fail when highly connected nodes are not connected directly to ideal voltage sources. One example of this is illustrated in Fig. 7, in which a parasitic network is attached to both VDD and VSS. Singleton removal depends upon the values being pre-determined and in the ex-

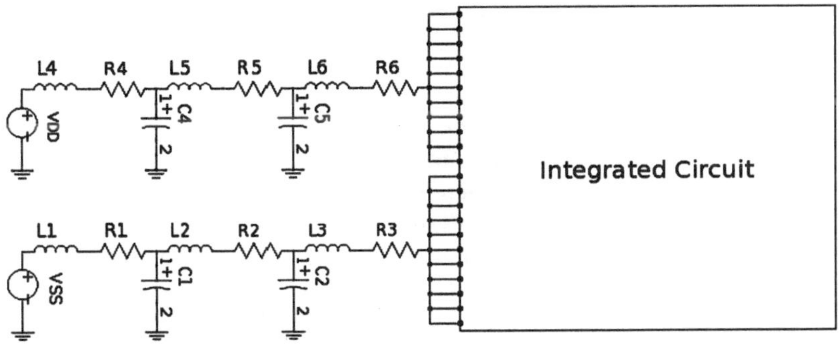

Fig. 7 Power node parasitics example. An RLC network sits between the VDD, VSS sources and the main circuit, so these highly connected nodes cannot be removed with a singleton filter

ample this will no longer be the case. As a result, both linear solution strategies illustrated in Fig. 5 would fail. Additionally, even if singleton removal did not fail, the additional structure created by the power node parasitic network is sufficient to destroy the triangular structure required by the BTF linear solver strategy.

Applying the multi-level Newton method can mitigate the graph problem presented by non-ideal power supplies. The power supply parasitic network on the left side of Fig. 7 is partitioned into one level, while the remaining integrated circuit on the right side is partitioned into another level. From a conceptual standpoint, either partition can be the top or bottom level, but it is more computationally efficient to put the parasitic network on the bottom level as it is (in this example) a relatively small circuit, while the integrated circuit comprises the bulk of the original problem. By using two-level partitioning, the values of VDD and VSS appear as ideal sources within the partition, and thus these matrix processing steps are preserved. A successful example of this approach is given in Sect. 7.

5 Software

Xyce [4] relies heavily on the open-source Trilinos scientific computing library [23], which was designed for parallel computing. Trilinos provides support for linear algebra data structures as well as preconditioners, linear solvers and nonlinear solvers. Xyce is not directly written to the Trilinos packages and, instead, has an abstract interface that defines the functionality necessary for performing circuit simulation. These software abstractions allow Xyce to use MPI with double precision arithmetic through Epetra or combine and adapt to other parallel paradigms.

Xyce is designed to use abstract interfaces wherever feasible for algorithmic components, so that the implementation of those components may be separated from the implementation of the simulator. Many benefits result from such a decision. This decoupling facilitates code reuse across Xyce and increases algorithmic flexibility. As a result, constituent mechanisms (e.g., nonlinear solvers, linear solvers, preconditioners) can be chosen at runtime. This enables Xyce to be a production simulator, as well as a testbed for parallel algorithm research.

Xyce uses abstract interfaces and runtime polymorphism throughout the simulation code. Much of the higher-level abstractions, relating to the analysis type or time integration methods, have implementations that are contained in Xyce. However, the lower-level abstractions, relating to nonlinear solvers, linear solvers, and basic linear algebra, have interfaces to Trilinos packages. Figure 8 shows the nested solver loop from the circuit simulation flow diagram (Fig. 1), where the shaded box illustrates the part of Xyce where Trilinos is utilized.

These software abstractions allow Xyce to adapt to future parallel paradigms and arithmetic precision strategies. By default, Xyce uses MPI with double precision arithmetic through Epetra. However, future computational platforms may require the use of other parallel paradigms, such as TBB [5] and CUDA [2], to achieve optimal performance. Furthermore, to address ill-conditioned matrices, it may prove useful to use quad-precision arithmetic in the linear solvers.

These forward-looking computational strategies are the motivation for the newer Trilinos linear algebra packages: templated Petra (Tpetra) and Kokkos. Tpetra provides a templated interface to parallel linear algebra and Kokkos contains the underlying computational kernels enabling platform-dependent optimizations. Several pre-existing Trilinos packages can use TPetra, like NOX, LOCA, Belos, and Teuchos, and many other packages are under development to provide direct solvers and preconditioners using Tpetra.

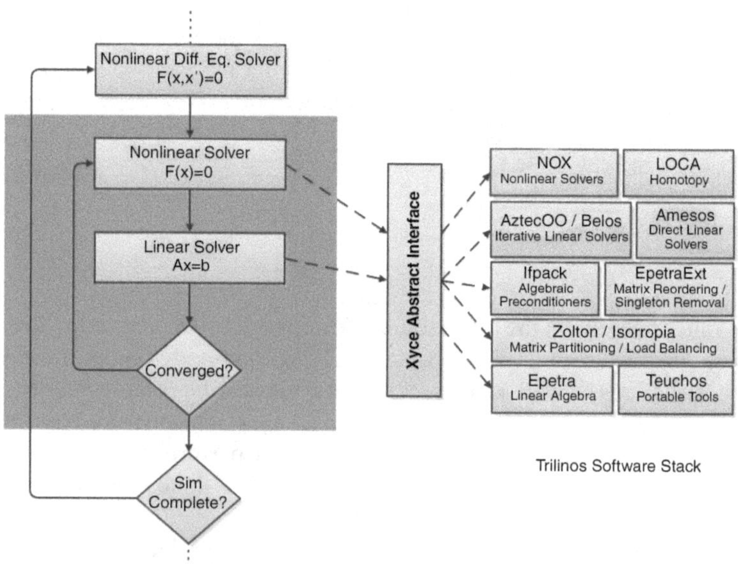

Fig. 8 Xyce simulation flow with interface to Trilinos

Table 1 Circuits: matrix size (N), capacitors (C), MOSFETs (M), resistors (R), voltage sources (V), diodes (D)

Circuit	N	C	M	R	V	D
ckt1	688838	93	222481	176	75	291761
ckt2	434749	161408	61054	276676	12	49986
ckt3	116247	52552	69085	76079	137	0
ckt4	46850	21548	18816	0	21	0
ckt5	25187	0	71097	0	264	0

6 Parallel Linear Solver Strategy Comparison

In this section, Xyce scalability experiments for several different circuits using various linear solvers are presented. The first comparison is an abbreviated version of the results presented in [16], comparing the domain decomposition (DD) and block triangular form (BTF) linear solver strategies with two other direct linear solvers: KLU and SuperLU_DIST (SLUD). KLU is a serial sparse direct linear solver developed specifically for circuit simulation [31]. SLUD is a general purpose (not circuit-specific) distributed memory sparse direct linear solver [22]. The second comparison illustrates the scalability of the two linear solver strategies over an increasing number of processors for the largest test circuit.

Since the goal of developing efficient, scalable linear solvers is to extend the simulation capability to very large ICs, test circuits were chosen that produced large (more than 10^4 unknowns) linear systems. Table 1 partially describes the these circuits, where ckt4 is the chip2 circuit from the well known test suite CircuitSim90 [7] and the others are proprietary integrated circuits.

All computations are performed on a cluster with 2.2 GHz AMD four-socket, quad-core processors with 32 GB DDR2 RAM and an Infiniband interconnect using the OFED software stack. Each node of the machine has a total of sixteen cores, and the user can request anywhere from one to sixteen cores per node. If less than sixteen cores per node are used, the memory is evenly divided between the cores, and more memory is available for each core. For the first comparison, two parallel configurations were considered, 4 cores and 16 cores, on one compute node. For the second comparison, the parallel configurations vary the number of processors per node (ppn) used as well as the number of compute nodes.

6.1 Explanation of Tables

For each circuit, results are presented for setup, device evaluation, solve and total times. The "setup time" refers to the time required to read the netlist, setup the circuit topology, allocate devices, resolve parameters, and create the linear system structure. The "device evaluation time" refers to the total time required to compute the Jacobian matrix and residual entries for all the individual devices, and to sum them into the parallel linear algebra structures, for every Newton iteration of every time step. The "solve time" is the total linear solve time for the entire transient simulation. The "total" time is simply the total time for the entire simulation, including setup, device evaluation, and solve times.

In Tables 2 and 3, there are a number of entries labeled F_1, F_2, and F_3 representing different failure modes for the tested solvers. The F_1 failure is unique to the BTF linear solver strategy, stemming from the block triangular form having one large irreducible block that makes it impossible to find a suitable partition. The F_2 failure results from Newton's method failing to converge, which ultimately leads to Xyce exiting with a time-step-too-small failure. The F_2 failure can happen with either iterative or direct solvers, but mostly happens with the iterative solvers when the preconditioner is ineffective. The F_3 failure mode denotes circuits that run out of memory, suggesting that the memory requirements of the solver are too high.

6.2 Numerical Results

KLU reliably solves all the test circuits, and is considered the baseline for both comparisons. Table 2 shows the simulation time for three of the five circuits, where the simulations using KLU are run on one core and the others are run on four cores. The BTF linear solver strategy successfully solves all three test circuits and the total simulation time has been reduced relative to serial KLU. The speedups are due to a combination of multiple cores and an improved algorithm, and can thus be superlinear. The DD linear solver strategy successfully solves two of the three test circuits, but is only faster than KLU on ckt5. Using the SLUD solver, the total simulation time is faster than KLU in two of the three test circuits.

Table 3 shows the simulation time for the three largest circuits on sixteen cores. The BTF linear solver strategy, when successful, achieves a substantial speedup and even superlinear speedup for ckt3. However the failure to solve ckt2 is due to its bidirectional structure, which results in a large irreducible block. The DD linear solver strategy does poorly on these larger test circuits, producing convergence failures in all but ckt1. SLUD also has difficulties, running out of memory on ckt1 and causing a convergence failure on ckt2.

The second comparison illustrates the scalability of the two linear solver strategies for ckt1. This comparison also examines the performance of loading values into the Jacobian matrix and the residual vector, which includes the device evaluation. The scaling is done relative to the serial simulation performance, where KLU is used as the linear solver. The simulations were run on 8, 16, 32, and 64 processors (cores) where 4 processors per node (ppn) were used. Thus 2, 4, 8, and 16 nodes were used to perform this study. The plot in Fig. 9 illustrates that the scaling of the Jacobian and residual load are about the same, as one would expect. However, the BTF linear solver strategy is almost twice as fast as the DD approach.

Table 2 Simulation times in seconds for four cores on a single node

Circuit	Task	KLU (serial)	SLUD	DD	BTF	**Speedup** (KLU/BTF)
ckt3	Setup	131	56	F_2	57	2.3x
	Device Eval	741	568	F_2	562	1.3x
	Solve	6699	2230	F_2	255	26.2x
	Total	7983	2903	F_2	923	8.6x
ckt4	Setup	15	8	8	13	1.2x
	Device Eval	1189	330	264	265	4.5x
	Solve	536	2540	1746	312	1.7x
	Total	1858	2916	2050	619	3.0x
ckt5	Setup	57	21	21	22	2.6x
	Device Eval	801	219	218	221	3.6x
	Solve	346	318	280	67	5.2x
	Total	1360	606	567	360	3.8x

Table 3 Simulation times in seconds for sixteen cores on a single node

Circuit	Task	KLU (serial)	SLUD	DD	BTF	**Speedup** (KLU/BTF)
ckt1	Setup	2396	F_3	207	199	12.0x
	Device Eval	2063	F_3	194	180	11.4x
	Solve	1674	F_3	3573	310	5.4x
	Total	6308	F_3	4001	717	8.8x
ckt2	Setup	2676	F_2	F_2	F_1	
	Device Eval	1247	F_2	F_2	F_1	
	Solve	1273	F_2	F_2	F_1	
	Total	5412	F_2	F_2	F_1	
ckt3	Setup	131	29	F_2	29	4.5x
	Device Eval	741	181	F_2	175	4.2x
	Solve	6699	1271	F_2	84	79.8x
	Total	7983	1470	F_2	306	26.1x

Relatedly, the BTF approach is a better choice than DD with respect to overall parallel scaling. When the DD strategy is used, the total simulation scaling falls midway between the Jacobian load and linear solve scalings. This indicates that the linear solve is a bottleneck to overall parallel performance, representing a larger fraction of the total runtime. However, with the BTF linear solver strategy, the total scaling is consistent with the Jacobian load scaling, demonstrating that the BTF-based linear solve does not impact the overall simulation scaling.

Figure 10 shows the total linear solve time for ckt1 over increasing numbers of processors, for three values of processors per node. This plot verifies the performance difference between the BTF and DD linear strategies, independent of the number of processors per node. Furthermore, it illustrates that increasing the number of processors past 32 is not likely to speed up the simulation. For any fixed problem size, this roll-off is to be expected beyond a certain number of processors. However, it should be noted that the overall runtime on 32 processors is approximately twenty times faster than the serial case, which is still a substantial improvement.

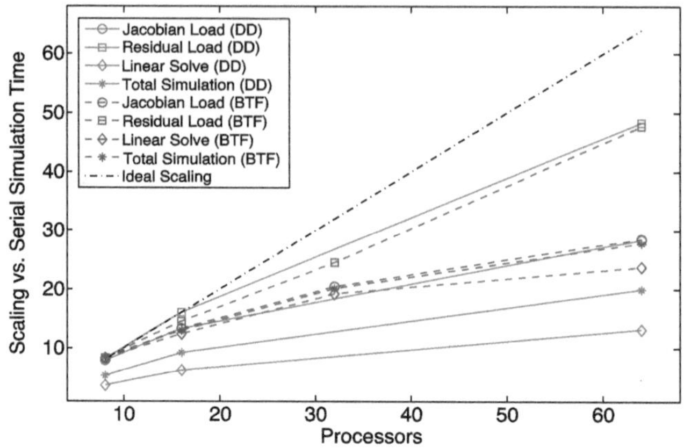

Fig. 9 Xyce scaling study for ckt1 using 4 ppn

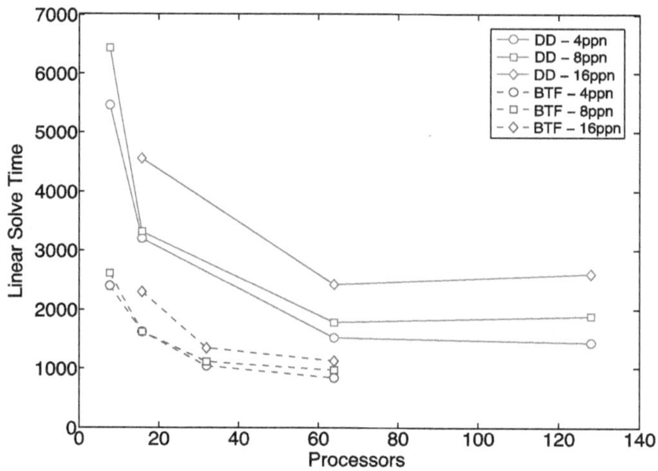

Fig. 10 Xyce linear solver scaling for ckt1 using 4, 8, and 16 ppn

7 Graph Mitigation Example

All the results given in the previous section are for circuits which do not have a parasitic network attached to highly connected nodes such as VDD and VSS (power and ground). However, if such a network were to be attached, both the DD and BTF solver strategies would fail. Here, an example is given, in which a power network similar to the one shown in Fig. 7 has been added to ckt3.

The initial consequences of adding the parasitic network are given in Fig. 11. In these figures, the DCOP matrix structure from the modified ckt3 is given, before and after the BTF reordering phase of the BTF solver strategy. The structure has not changed significantly, and the reordered matrix has a single irreducible block. As a

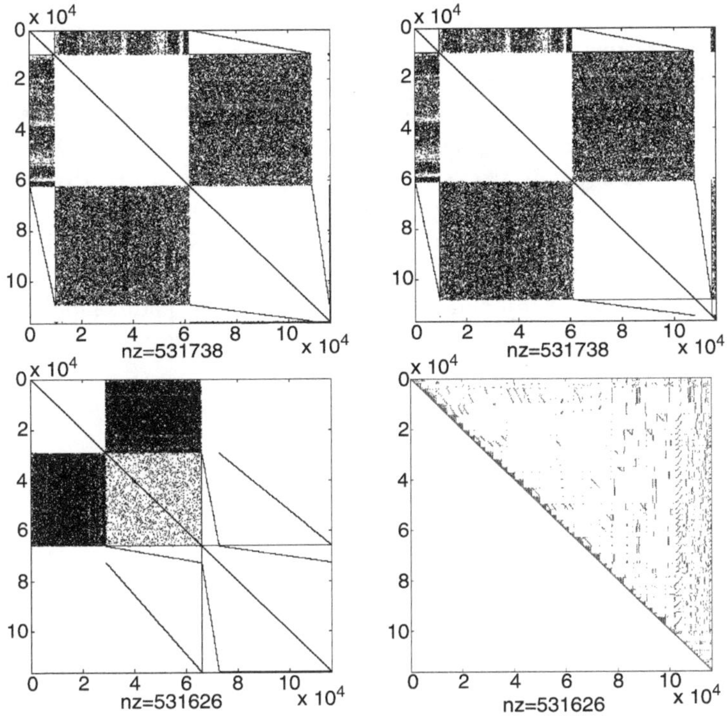

Fig. 11 Matrix pattern for ckt3 with power node parasitics attached before (top, left) and after (top, right) BTF reordering. After the circuit has been partitioned into a multi-level Newton solve, the matrix pattern for the inner solve before (bottom, left) and after (bottom, right) BTF reordering

result, it is not possible to achieve an effective block-wise load balance in parallel, as the irreducible block will be confined to a single processor.

The graph problems presented by the parasitic network can be mitigated by using the multi-level Newton approach described in Sect. 4. The parasitic network (corresponding to the left side of Fig. 7) in this example is considered the upper level of the problem, while the rest of the integrated circuit is considered to be the lower level. The graph problems are thus confined to the upper level and invisible to the lower. The result of this strategy is given in Fig. 11.

8 Conclusion

Parallel circuit simulation requires integration of large and small scale parallelism throughout the entire circuit simulation flow. In this paper, parallel algorithms for circuit simulation, and their implementation in a production simulator, Xyce, have been discussed. Specific attention was given to parallelism issues in the netlist parser, nonlinear solver, and linear solver.

A key design aspect for Xyce is the development of new circuit-specific preconditioners for iterative linear solvers. While not as robust as direct solvers, iterative solver strategies have the potential to enable scalable parallel simulation. The results presented show that the BTF linear solver strategy reduces the total simulation time by up to a factor of twenty compared to the serial solver KLU on 32 processors. However, the strategy only works well when the conductance matrix is reducible, which is not true for some circuits for a variety of different reasons. One graph mitigation strategy, multi-level Newton, has been successful at partitioning an irreducible graph to a reducible one. Future work will include a focus on developing other strategies for graph mitigation. Ultimately, a general parallel tool will require a comprehensive strategy to obtain good parallel performance.

Acknowledgements The authors would like to thank Mike Heroux, Ray Tuminaro, David Day, Erik Boman, and Scott Hutchinson for many helpful discussions.

References

1. *Cadence UltraSim.* `http://www.cadence.com/products/cic/UltraSim_fullchip/`.
2. *NVIDIA CUDA programming guide.* `http://www.nvidia.com/object/cuda.html`.
3. *Synopsys HSIM.* `http://www.synopsys.com/Tools/Verification/AMSVerification/CircuitSimulation/HSIM/`.
4. *Xyce Parallel Circuit Simulator.* `http://xyce.sandia.gov`.
5. *Intel Threading Building Blocks 2.0.* `http://www.intel.com/software/products/tbb/`, March 2008.
6. P. Amestoy, T. Davis, and I. Duff. An approximate minimum degree ordering algorithm. *SIAM J. Matrix Anal. Appl.*, 17(4):886–905, 1996.
7. J. Barby and R. Guindi. Circuitsim93: A circuit simulator benchmarking methodology case study. In *Proceedings Sixth Annual IEEE International ASIC Conference and Exhibit*, Rochester NY, 1993.
8. A. Basermann, U. Jaekel, M. Nordhausen, and K. Hachiya. Parallel iterative solvers for sparse linear systems in circuit simulation, January 2005.
9. C. Bomhof and H. van der Vorst. A parallel linear system solver for circuit simulation problems. *Num. Lin. Alg. Appl.*, 7(7-8):649–665, 2000.
10. T. A. Davis. *Direct Methods for Sparse Linear System.* SIAM, 2006.
11. D. M. Day, M. K. Bhardwaj, G. M. Reese, and J. S. Peery. Mechanism free domain decomposition. *Comput. Meth. Appl. Mech. Engrg.*, 182(7):763–776, 2005.
12. K. Devine, E. Boman, R. Heaphy, R. Bisseling, and U. Catalyurek. Parallel hypergraph partitioning for scientific computing. In *Proceedings of 20th International Parallel and Distributed Processing Symposium (IPDPS'06)*. IEEE, 2006.
13. N. Fröhlich, B. Riess, U. Wever, and Q. Zheng. A new approach for parallel simulation of vlsi-circuits on a transistor level. *IEEE Transactions on Circuits and Systems Part I*, 45(6):601–613, 1998.
14. W. Gropp, E. Lusk, N. Doss, and A. Skjellum. A high-performance, portable implementation of the MPI message passing interface standard. *Parallel Computing*, 22(6):789–828, September 1996.

15. K. Gulati, J. F. Croix, S. P. Khatr, and R. Shastry. Fast circuit simulation on graphics processing units. In *Proceedings of the 2009 Conference on Asia and South Pacific Design Automation*, pages 403–408. IEEE Press, 2009.
16. H. K. Thornquist et al. A parallel preconditioning strategy for efficient transistor-level circuit simulation. In *Proceedings of the 2009 International Conference on Computer-Aided Design*. ACM, 2009.
17. B. Hendrickson and R. Leland. A multilevel algorithm for partitioning graphs. In *Proceedings of Supercomputing '95*. ACM, December 1995.
18. G. Karypis, R. Aggarwal, V. Kumar, and S. Shekhar. Multilevel hypergraph partitioning: Applications in VLSI domain. *IEEE Trans. VLSI Systems*, 20(1), 1999.
19. G. Karypis and V. Kumar. ParMETIS: Parallel graph partitioning and sparse matrix ordering library. Technical Report 97-060, CS Dept., Univ. Minn., 1997.
20. B. W. Kernighan and S. Lin. An efficient heuristic procedure for partitioning graphs. *The Bell System Technical Journal*, 49(2):291–307, 1970.
21. K. Kundert. Sparse matrix techniques. In A. Ruehli, editor, *Circuit Analysis, Simulation and Design*. North-Holland, 1987.
22. X. S. Li and J. W. Demmel. SuperLU_DIST: A scalable distributed-memory sparse direct solver for unsymmetric linear systems. *ACM Trans. Math. Softw.*, 29(2):110–140, 2003.
23. M. Heroux, et al. An overview of the Trilinos project. *ACM TOMS*, 31(3):397–423, 2005. `http://trilinos.sandia.gov`.
24. F. Manne and R. H. Bisseling. A parallel approximation algorithm for the weighted maximum matching problem. In *Proceedings of PMAA'08, LNCS 4967*, pages 708–717. Springer, Berlin, 2008.
25. W. McLendon, B. Hendrickson, S. Plimpton, and L. Rauchwerger. Finding strongly connected components in distributed graphs. *J. of Parallel and Distributed Computing*, 65:901–910, 2005.
26. L. W. Nagel. Spice 2, a computer program to simulate semiconductor circuits. Technical Report Memorandum ERL-M250, 1975.
27. A. R. Newton and A. L. Sangiovanni-Vincentelli. Relaxation based electrical simulation. *IEEE Trans. Comput.-Aided Design Integr. Circuits Syst.*, CAD-3(4):308–330, October 1984.
28. H. Peng and C.-K. Cheng. Parallel transistor level circuit simulation using domain decomposition methods. *ASP-DAC '09: Proceedings of the 2009 Conference on Asia and South Pacific Design Automation*, pages 397–402, Jan 2009.
29. N. B. G. Rabbat, A. L. Sangiovanni-Vincentelli, and H. Y. Hsieh. A multilevel Newton algorithm with macromodeling and latency for the analysis of large-scale nonlinear circuits in the time domain. *IEEE Trans. Circuits Syst.*, CAS-26(9):733–741, 1979.
30. Y. Saad. *Iterative Methods for Sparse Linear Systems*. SIAM, Philadelphia, PA, second edition, 2003.
31. K. Stanley and T. Davis. KLU: a 'Clark Kent' sparse LU factorization algorithm for circuit matrices. In *SIAM Conference on Parallel Processing for Scientific Computing (PP04)*, 2004.
32. K. Sun, Q. Zhou, K. Mohanram, and D. C. Sorensen. Parallel domain decomposition for simulation of large-scale power grids. *IEEE/ACM International Conference on Computer-Aided Design (ICCAD)*, pages 54–59, November 2007.
33. B. Ucar and C. Aykanat. Partitioning sparse matrices for parallel preconditioned iterative methods. *SIAM J. Sci. Comput.*, 29(4):1683–1709, 2007.
34. J. White and A. L. Sangiovanni-Vincentelli. Relax2: A new waveform relaxation approach for the analysis of lsi mos circuits. In *Proc. Int. Symp. Circuits Syst.*, 2:756–759, May 1983.

A Perspective on Fast-SPICE Simulation Technology

Michał Rewieński

Abstract This chapter presents an introduction to the area of accelerated transistor-level ('fast-SPICE') simulation for automated verification and characterization of integrated circuits (ICs) from technologist's perspective. It starts with outlining goals, expectations and typical usage models for fast-SPICE simulators, stressing how they differ from regular SPICE tools. It continues with presenting and classifying core technologies typically included in fast-SPICE simulators, which allow them to achieve critical performance and capacity gains. Also, it discusses how different approaches toward the computational problem can be combined to design and implement highly effective and efficient simulation acceleration technologies, including advanced circuit partitioning, and schemes for optimized modeling of post-layout memory circuits and large parasitic networks. Finally, challenges facing fast-SPICE, as well as possible future research areas are briefly outlined.

1 Introduction

Transistor-level electrical simulation of integrated circuits, i.e. SPICE simulation, is a dynamic area of academic and industrial research. Throughout the years the accompanied software development provided electrical engineers with vital tools and methodologies allowing them to characterize and verify new designs prior to their fabrication in silicon. SPICE simulation quickly became an inherent component of the integrated circuit (IC) design process, crucial to successful and cost-efficient production of electronic chips. Naturally, SPICE tools became part of the race to develop increasingly more complex circuits, and were required to match their simulation capabilities with rapidly advancing transistor manufacturing technology (and conversely, it became clear that capacity of simulation software puts a limit on com-

Michał Rewieński
Synopsys Inc., 700 E. Middlefield Rd., Mountain View, CA, USA
e-mail: michalr@synopsys.com

P. Li et al. (eds.), *Simulation and Verification of Electronic and Biological Systems*,
DOI 10.1007/978-94-007-0149-6_2, © Springer Science+Business Media B.V. 2011

plexity of ICs which can be efficiently produced). Further integration and downscaling of electronic circuits created a need to simulate larger design blocks, as well as entire chips. This meant scaling up simulation capabilities from circuits with thousands of transistors to circuits with millions of transistors, which, as quickly realized, required abandoning the uncompromising approach to simulation accuracy of the initial SPICE tools. Hence, a new class of *accelerated* transistor-level simulators (also called 'fast-SPICE' simulators) emerged. 'Fast-SPICE' tools started to explore various accuracy tradeoffs to achieve incredible performance and capacity gains. Those in turn were made possible by a myriad of novel heuristics, methodologies, and computational techniques.

Discussing fast-SPICE technologies is the focus of this chapter. Yet, presenting a comprehensive summary or even a listing of all the simulation acceleration methods, published in hundreds of papers and patents, lies well beyond the scope of this contribution. Instead, this chapter attempts to take only a perspective on those technologies. The perspective will be that of a fast-SPICE researcher and developer at the same time (which is the author's own), and will allow one to see the inherently multidisciplinary character of fast-SPICE technologies, underline some of the common characteristics of certain successful techniques, and at the same time illustrate general shifts of focus of fast-SPICE simulation over the years.

First, the scene is set up by briefly presenting the computational problem shared by SPICE and fast-SPICE, and the initial approach toward solving it, exemplified by virtually all classical SPICE simulators. Then, typical usage and limitations of SPICE simulators are outlined, followed by motivations behind accelerated, fast-SPICE tools, and presentation of main differences between the two classes of tools. Next, fast-SPICE technologies are discussed with emphasis on their various, often complementary views of the computational problem, and their classification is attempted. A few selected techniques are then described in more detail in order to illustrate how they integrate various approaches and views of the computational problem in order to achieve gains in simulation capacity and performance. Finally, challenges and potential research paths in fast-SPICE simulation are listed.

2 SPICE: Transistor-Level Circuit Simulation

In SPICE simulation transistor models (such as e.g. BSIM4 or Mos11), lumped passive element models (e.g. R, L, C), and voltage and current source models (e.g. using modified nodal analysis (MNA) formulation) are combined with circuit equations (Kirchhoff's current and voltage laws) to describe the physics and topology of the considered integrated circuit design. In mathematical terms, the problem to be solved numerically is a system of differential-algebraic equations (DAEs) in the following form:

$$\frac{dq(\mathbf{v},t)}{dt} = i(\mathbf{v},\mathbf{u},t) \,, \tag{1}$$

where $\mathbf{v} = \mathbf{v}(t)$ is a vector containing nodal voltages and branch currents (for elements applying MNA formulation), $\mathbf{u} = \mathbf{u}(t)$ is a vector of inputs (realized using e.g. voltage sources or digital vector input elements), t is time, $q()$ is an operator describing charge, and $i()$ is an operator describing current. System (1) modeling an integrated circuit is typically highly nonlinear, stiff and, in the case of most fast-SPICE applications, has an extremely large size. Most commonly one is interested in finding transient time-domain solution of (1) in response to a given input signal, therefore the following description will focus on this case.

In order to tackle problem (1) numerically one needs to address the following issues:

1. A numerical integration scheme, along with suitable corresponding time discretization (time-stepping) needs to be applied. Popular and inexpensive choices for stiff systems include backward-Euler, trapezoidal, and second-order Gear methods. For instance, backward-Euler integration will yield an algebraic equation in the form shown below:

$$\mathbf{v}_{t+\Delta t} = \mathbf{v}_t + \Delta t f(\mathbf{v}_{t+\Delta t}, \mathbf{u}_{t+\Delta t}, t + \Delta t) \tag{2}$$

   where $\mathbf{v}_t$ is a (known) vector of nodal voltages at time t, $\mathbf{v}_{t+\Delta t}$ is an unknown vector of nodal voltages at time $t + \Delta t$, Δt is the time-step, $\mathbf{u}_{t+\Delta t}$ is a vector of inputs at time $t + \Delta t$, and $f()$ is a (nonlinear) operator. In the above formulation, the unknown vector $\mathbf{v}_{t+\Delta t}$ is not given explicitly in (2), and needs to solved for.
2. Algebraic equation (2) needs to be solved in order to find $\mathbf{v}_{t+\Delta t}$. It can be rewritten as a problem of finding zeros of a nonlinear operator:

$$F(\mathbf{v}) = 0 \tag{3}$$

   where $\mathbf{v}$ is the unknown vector of nodal voltages at time $t + \Delta t$, and $F(\mathbf{v}) = \mathbf{v} - \mathbf{v}_t - \Delta t f(\mathbf{v}, \mathbf{u}_{t+\Delta t}, t + \Delta t)$. Zeros of the above equation are most often found using some flavor of the multi-dimensional Newton-Raphson method, yet other choices are available such as e.g. successive chords method [8, 16]. An iteration of the Newton method for (3) is given below:

$$\mathbf{v}^{i+1} = \mathbf{v}^i - J_F^{-1}(\mathbf{v}^i) F(\mathbf{v}^i) \tag{4}$$

   where J_F is the Jacobian matrix for operator F, $\mathbf{v}^i$ is an approximation of $\mathbf{v}$ at i-th iteration, $\mathbf{v}^{i+1}$ is approximation of $\mathbf{v}$ at $(i+1)$-th iteration.
3. Finally, a system of linear equations needs to be solved at each iteration of the Newton process (4):

$$J_F(\mathbf{v}_i)\mathbf{x} = F(\mathbf{v}_i). \tag{5}$$

   To this end any direct or iterative linear solver can be applied. In SPICE domain the 'default,' or reference choice of the linear solver is still a direct solver based on LU factorization with an appropriate pivoting scheme.

The above mentioned components: a numerical integration method with a time-stepping scheme suitable for stiff systems of DAEs, robust nonlinear and linear solvers, along with detailed, rigorous device models (specified in terms of mathematical equations relating physical quantities such as voltages and currents, as well as their derivatives) constitute the core of a SPICE simulation engine. This view of the computational engine heavily influenced design of SPICE simulators which were typically implemented as a combination of interacting yet separate numerical components.

2.1 Usage and Limitations of SPICE

The mathematical and numerical framework of SPICE, very briefly outlined in the previous paragraph, has been set up with one main goal in mind: to provide means for high precision electrical modeling of integrated circuits. SPICE tools based on this framework allowed engineers to simulate their design with the highest achievable accuracy. Therefore, SPICE tools naturally dominated the following simulation/modeling areas:

- Characterization of cell libraries and Intellectual Property (IP) circuit blocks;
- Modeling of precision analog components;
- Verification of subsystems along critical paths.

Since achieving highest possible accuracy of the results was a priority (aligned with expectations of SPICE users), a number of restrictions were imposed on the simulation engine. Above all those included: no approximations to device models, uniform time discretization across the entire simulated circuit, and global convergence to the solution at every time-point. This in turn meant constructing and solving a single system of equations representing the entire circuit at every time-point. This synchronous solve often implied taking very small steps even for very latent circuit blocks. The need to solve a single large and stiff linear system, typically meant that direct linear solvers were the only class of methods which could provide reliable performance. (Finding robust preconditioners for iterative methods often proved to be extremely challenging, hence drastically limiting usability for this class of solvers.)

Yet, using direct (sparse) solvers also meant that the cost of solving the linear system was reaching $O(N^{1.5})$ or even $O(N^2)$ for some circuit topologies, where N was the number of nodes in a circuit. Apart from the cost of evaluating complex, equation-based device models (e.g. for MOSFETs), the cost of linear solves remains a critical factor limiting capacity of SPICE simulators. Although this capacity limits for full-accuracy SPICE tools is constantly pushed upwards, its current rough estimate is a circuit with about 100,000 active elements (MOSFETs).

2.2 Need for Accelerated SPICE ('Fast-SPICE') Simulation

Higher levels of integration, and shrinking IC manufacturing process technology created a need to simulate and verify at transistor level much larger circuit blocks, well beyond capacity limits of conventional SPICE simulators. In particular two important simulation areas emerged:

- Verification of large-scale digital, memory circuits;
- Mixed-signal, system-on-a-chip (SoC), full-chip functional simulation.

In order to be able to perform such simulations, involving designs with active element count reaching from 100 million to 1 billion MOSFET devices, and millions of parasitic devices, the capacity of SPICE tools had to increase dramatically. For problems of this size the $O(N^{1.5})$ complexity of linear solves became clearly unacceptable, creating a need to modify or abandon some of the paradigms followed by conventional SPICE simulators, such as e.g. single-matrix solve for global solution convergence. At the same time, accuracy requirements for full-chip functional simulations or memory verification runs were clearly not as stringent as for e.g. numerical characterization of cell libraries, allowing for 5–15% error as compared to full SPICE accuracy. IC designers were willing to pay a price for the ability to simulate much larger systems. This created an excellent opportunity to explore various tradeoffs between achievable accuracy, and simulator speed and capacity. Hence, both the need for transistor-level simulations for much larger systems, and more moderate expectations on accuracy for large-scale simulations drove development of a new class of tools: the accelerated transistor-level ('fast-SPICE') simulators.

3 Fast-SPICE Technologies

Relaxed accuracy constraints on large-scale simulations enabled development of simulation optimization technologies which by large and far surpassed optimizations considered in traditional SPICE tools, yielding tremendous performance gains reaching 1,000- or 10,000-fold speedups, as well as similar capacity improvements. Hence, it became feasible to perform simulations for circuits with hundreds of millions of devices in a matter of minutes—as illustrated in Table 1, which shows sample performance data for a commercial fast-SPICE tool.

3.1 SPICE vs. Fast-SPICE Techniques

Key fast-SPICE technologies developed to achieve such performance gains differed fundamentally from optimizations found in conventional SPICE engines. In SPICE tools, the only simulation optimizations permitted were those which did not cause any significant loss of accuracy (cf. Fig. 1). Furthermore, due to the assumption that

Table 1 Performance of a commercial fast-SPICE simulator for a sample circuit containing 284 million active MOSFET devices, 290 million passive RC elements, represented using a hierarchical netlist. The test was performed on a workstation with 2.8GHz AMD Opteron processor, using 64-bit precision

	CPU time	Memory cost
Netlist parsing	0 s	18 MB
Identification, optimization	215 s	853 MB
Operating point computation	70 s	395 MB
Transient simulation	674 s	538 MB
Total	959 s	1804 MB

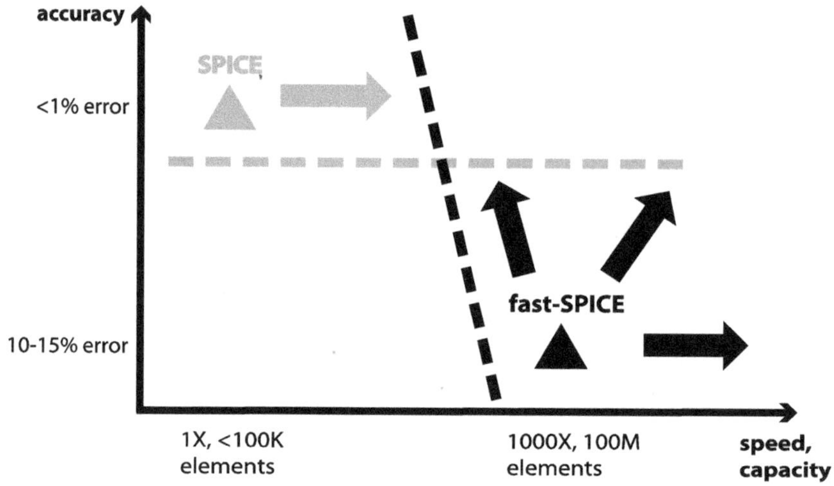

Fig. 1 Triangles show typical 'operating points' for SPICE and fast-SPICE in terms of their capacity, performance and accuracy of computed results. The only allowed SPICE optimizations (indicated with a gray arrow) are those which increase its capacity and speed without sacrificing its high accuracy. On the contrary, in fast-SPICE simulation optimization technologies are allowed to explore various speed/capacity vs. accuracy tradeoffs (cf. black arrows), providing the user with flexibility to select target accuracy or degree of simulation acceleration

SPICE should be able to accurately simulate *any* design, including any novel circuit topology, or unconventional design etc., relatively little effort has been dedicated toward identifying *typical* functional circuit components or blocks, which could potentially be simulated more efficiently. Consequently, the starting point for efforts to improve SPICE was typically a rather narrow view of the computational problem as predominantly a system of equations to be solved. This in turn implied that improvements focused on optimizing components of SPICE engine's mathematical machinery. Enhancements were most often made separately for time-integration schemes, nonlinear solvers, and linear solvers. While over time significant progress has been achieved, yielding gains both in speed and capacity of SPICE tools, the process of improving SPICE in such a manner proved to be very challenging and relatively slow.

During development of fast-SPICE technologies a very different approach toward optimizing the computational problem has been taken. First of all, as already mentioned, those optimizations were allowed to trade off simulation accuracy for speed and capacity, and also permitted direct or indirect control over the desired target level of accuracy (cf. Fig. 1). A no less important consideration was that fast-SPICE tools were to be predominantly used for verification of large systems composed of relatively conventional and robust functional components. This implied that simulation could be aggressively optimized for many well understood circuit blocks. This new approach toward optimizing the simulation dramatically broadened the viewpoint at the computational problem at hand. It was no longer just a system of nonlinear differential equations which was to be optimized, but it became a system of equations describing often well identified functional blocks with known behavior and physical properties. Consequently, instead of narrowly focused SPICE optimizations, fast-SPICE tools contained technologies which took a more *holistic* approach towards optimizing the simulation problem, which was often key to their success. As it will be illustrated in the following sections, fast-SPICE techniques applied, sometimes simultaneously, very different viewpoints in order to achieve tremendous performance and capacity gains.

As a starting point for this discussion let us briefly mention a few 'signature' fast-SPICE optimizations. Then we will go deeper into explaining some general approaches which stimulated their development. The key fast-SPICE technologies are listed in Table 2, which also compares them to solutions used in conventional SPICE. Firstly, accurate, equation-based models for MOSFETs are approximated in fast-SPICE by simplified equations or, most notably, tabularized models for device current, capacitance and charge, yielding 1,000-fold speedups. For passive elements model order reduction (MOR) techniques are applied in order to generate compact models for interconnects. Fast-SPICE also abandoned the requirement for global convergence at each time-point, and global uniform time-stepping for the entire circuit. Instead, 'event-driven' simulation is performed, where solution in a given region is computed only if a change of state is anticipated due to a changing input or load for this region. Also, instead of using global uniform time-step for the entire circuit, fast-SPICE performs 'multi-rate' simulation, where depending on the rate of nodal voltage change in different circuit regions, as well as other criteria, different time-steps are used during numeric integration. Last, but not least, fast-SPICE no longer solves a single system of equations, with a corresponding single, large ma-

Table 2 SPICE vs. fast-SPICE: key optimizations

SPICE	Fast-SPICE
Accurate device models	Approximate MOSFET models, reduction of parasitic networks
Global convergence at each time-point	Event-driven simulation, no global convergence
Global time-step	Multi-rate simulation
Single matrix for entire circuit	Partitioned matrix, hierarchical simulation

trix used during linear solves. Instead, the problem is partitioned into a number of blocks, using carefully designed physical and/or mathematical criteria. Then, linear systems (with corresponding matrices) are solved separately for each block. The final solution is found by performing e.g. some form of relaxation across different blocks, if needed.

All the techniques mentioned above cause systematic accuracy loss, and are therefore specific to fast-SPICE. Apart from them, there were a number of important optimizations initially developed in fast-SPICE domain, which provide performance gains without sacrificing simulation accuracy. Those include:

- High-capacity, high-performance input data storage and processing algorithms for simulation setup phase (i.e. simulator 'front-end'), e.g. efficient methods for processing highly hierarchical netlist representation of the simulated circuit.
- High performance linear system solve technologies, including efficient preconditioners and iterative algorithms, as well as parallel and multi-threaded solvers.
- Automatic, on-the-fly code generation and compilation for faster execution of key simulation blocks.

Although the above mentioned techniques are not fast-SPICE specific, many of them emerged during fast-SPICE development, due to a pressing need for efficient processing of very large input data sets such as hierarchical and non-hierarchical input netlists for full-chip simulations, as well as corresponding back-annotation files, generated by RC extraction tools for post-layout simulation. Many of these software and algorithmic techniques were later transferred to conventional SPICE simulators to increase their capacity and front-end performance.

3.2 Acceleration Technologies: Different Viewpoints

In order to gain more insight on how successful technologies for accelerating transistor-level simulation are designed one should take a look at the general methodology applied during fast-SPICE development, as well as different viewpoints at the computational problem which provide starting points for algorithmic innovations.

3.2.1 Fast-SPICE as Knowledge Database

First of all, fast-SPICE simulators are designed to be much more than conglomerates of various numerical techniques for solving large systems of stiff DAEs. Rather, they are developed as knowledge databases, in which expertise on design, structure, and functionality for many existing IC designs is heavily exploited in order to accelerate simulation while controlling accuracy of the results. Fast-SPICE tools typically include methods which automatically or semi-automatically identify functional circuit blocks of different complexities in order to split the computational problem and to use appropriate solving and modeling techniques and approaches for each block. Also, specialized techniques (some of which will be described in the next section) are implemented in fast-SPICE for certain large-scale functional blocks with well-known functionality and characteristics (such as e.g. power supply networks or memory circuits) in order to achieve cutting-edge simulation speed.

While examining fast-SPICE algorithms, one may distinguish a few complementary viewpoints at the computational problem. Those viewpoints or approaches provide different starting points for simulation acceleration technologies and may be classified as:

- Circuit-based: In this approach knowledge of electrical properties, and functional characteristics for circuit subsystems allows one to develop appropriate optimization and simulation strategy for each block.
- Graph-based: Treating the simulated circuit as a graph allows one to apply high-performance data representation and processing algorithms, as well as efficient identification of blocks with specific topologies.
- Matrix-based: Since, after all, a system of equation is solved, this viewpoint allows one to focus on optimizing matrix representation of the (linearized) problem in order to obtain a system of equations which is easier and faster to solve.

Next, we will mention various techniques which use each of the viewpoints listed above.

3.2.2 Matrix-Based Viewpoint

The 'matrix-based' approach has provided a starting point for many methods for accelerating transistor-level simulations. One group of such techniques, analyzes fill-in patterns for matrices representing linearizations of the system describing the considered circuit. Those techniques include:

- Node ordering schemes applied during matrix factorization and linear solves, e.g. minimum degree algorithm, nested dissection (cf. hMETIS [5]) designed to optimize matrix fill-ins;
- Matrix preconditioning techniques, including e.g. incomplete LU or multi-grid preconditioning [12];
- Fill-in based matrix partitioning (cf. Sect. 4.2);
- Parallelization techniques for linear solvers;
- Algorithms for selecting a linear solver (direct or iterative).

The 'matrix-based' approach is also applied in many leading Model Order Reduction (MOR) methods for constant positive real submatrices of the system matrix (corresponding to passive, linear portions of the circuit). Such MOR algorithms include Krylov subspace methods [2], or truncated balanced realization [10]. Finally, matrix approach is used in weak fill-in elimination, e.g. in techniques which compute Schur complement to discover weak couplings.

3.2.3 Graph-Based Viewpoint

On the other hand, the 'graph-based' approach stimulated development of the optimization technologies such as:

- Hierarchical circuit data processing and storage, leading to vast improvements in performance and capacity, predominantly in the simulation setup phase.
- Isomorphic matching techniques applied during transient simulation phase, which avoid repeated simulation of sufficiently identical circuit blocks ([1, 15]).
- Effective and efficient methods for topological identification of structure of circuit blocks (such as e.g. memory arrays), leading to methodologies such as hierarchical array reduction (cf. [7], Sect. 4.3).

It should also be mentioned that graph *concepts* are heavily exploited in many other fast-SPICE techniques, such as e.g. minimum degree or graph dissection-based node ordering schemes, as well as MOR methods already discussed above.

3.2.4 Circuit-Based Viewpoint

Last but not least, the 'circuit-based' approach has been used to develop a broad spectrum of simulation optimization techniques. Firstly, automatic or semi-automatic identification, and basic characterization of functional blocks for the simulated circuit allows one to apply optimized modeling approaches to specific block components. In particular:

- For transistors in a block which are found to be operating in digital regime, high performance approximate models (e.g. based on lookup tables) can be applied instead of full-precision equation-based models;
- For entirely linear circuit sub-blocks, model order reduction techniques can be used to generate efficient macromodels;
- Different integration, preconditioning, and solve algorithms can be applied to signal blocks either identified as digital or analog.

Also, the physical or circuit point of view allows one to effectively partition the computational problem. For problem partitioning purposes, fast-SPICE tools most often exploit knowledge on CMOS-based designs. For instance, identifying MOSFET channel connected blocks provides the very simplest partitioning technique. Once functional partitioning is done, sub-systems corresponding to different blocks can be solved separately, different integration time-points and time-steps can be used in each block, and achieving global convergence across different block may be optional. Also, solves for latent blocks may be avoided altogether.

Above we have classified various techniques for accelerated transistor-level simulation by the dominant approach (circuit-, graph-, or matrix-based) towards the computational problem. In fact, highly successful technologies most often *simultaneously* apply all the three discussed views. Next, we will present examples of state-of-the-art fast-SPICE technologies which from the start were developed with such holistic approach in mind.

4 Examples of Fast-SPICE Technologies

This section illustrates how different viewpoints at the simulation problem discussed above are exploited and combined in some advanced fast-SPICE technologies. The selected examples of computational techniques reach beyond the usual methods implemented in the initial fast-SPICE tools, such as methods for generating table transistor models, or simple problem partitioning based on MOSFET channel connected blocks (cf. e.g. [3]). Outlined are comprehensive techniques including an approach for accelerating simulation of parasitic networks, an advanced partitioning methodology, as well as a technique for accelerating simulation of memory circuits. All three technologies not only exemplify the holistic approach to the simulation problem, but also illustrate a general shift of focus in fast-SPICE from MOSFET-centric optimization methods, to acceleration techniques for problems which include simulation of coupling and other parasitic effects at a massive scale. Such effects challenge many of the simple heuristic acceleration techniques applied in older fast-SPICE tools, and point toward broader optimization strategies.

4.1 Optimized Simulation of Parasitic Networks

Fast-SPICE tools have applied various techniques for optimized simulation of passive RC and RLC networks. Those techniques have significantly evolved as parasitic effects became more and more pronounced, and as sizes of extracted R(L)C networks exploded.

Initially, relatively simple reduction techniques using the 'circuit-based' view of the RC network were developed. Those techniques applied simple transformations such as splitting floating capacitors, or shifting ground capacitors from one node to another in order to eliminate nodes [14]. Simple physical criteria (e.g. estimated signal delay, relative capacitor value) were used to decide if a transformation could be performed (cf. e.g. [13]).

Later, more advanced techniques exploiting the matrix (or state-space) view of the problem were applied in fast-SPICE in order to reduce the number of nodes in RLC networks. Those included various methods which used state-space projections to generate a more compact matrix for the linear (RLC) network, e.g. Krylov subspace methods, projections based on selecting dominant poles, truncated balanced realization etc. [2, 6, 10]. Such matrix-based techniques were often able to compress linear networks more aggressively than physical, circuit-based approaches.

Yet, both classes of algorithms mentioned above focused their efforts on reducing the number of nodes in a linear network (i.e. model order reduction). Such narrow focus, as well as interpreting the problem only in terms of either an electrical circuit or a state-space system, often rendered those methods unable to generate compressed networks which would result in faster simulations. In particular, matrix-based MOR algorithms at times yielded disastrous results for large linear networks, producing reduced order models which were dramatically slowing down simulations instead of speeding them up [11].

Nevertheless, broader look at the computational problem enabled further advancement of model order reduction (MOR) methods. For one class of recently developed successful MOR techniques (an example of such method is presented in [11]), the focus is shifted from reducing the number of nodes in a linear network to minimizing the number of fill-ins produced during matrix factorization, which is directly connected to the cost of linear solves. This is achieved by incorporating information on factorization into RC network node reduction process. For instance, method [11] employs the same node ordering algorithm during node reduction (when it decides whether a node should be eliminated or not), and during matrix factorizations. The cited method heavily relies on a graph view of the linear network to be reduced, using such graph theoretic concepts as articulation points, as well as approximate minimum degree (AMD) node ordering for reduction. Such comprehensive approach which integrates model order reduction, a graph-based view of the problem, as well as a matrix-based view associated with solving of the resulting linear system, leads to a method capable of creating efficient compressed models for large-scale linear networks with a huge number of driving terminals, yielding large speedups during simulation phase. Also, the applied graph-based approach resulted in a reduction method which is highly scalable, as well as on its own computationally inexpensive to execute. This is particularly important in fast-SPICE applications where potential for reuse of reduced models is very limited.

4.2 Advanced Partitioning Technologies

As already mentioned in previous sections, another key fast-SPICE feature, is partitioning of the computational problem into subproblems (with commensurate partitioning of the corresponding system matrix into submatrices). Efficient partitioning substantially reduces the cost associated with linear solve of a single system, and thereby makes simulation of circuits with tens of millions of nodes tractable.

Early fast-SPICE incarnations most often used the concept of strong feedforward i.e. weak dependence of voltage (e.g. of a digital gate or a voltage source) on the current through a MOSFET gate (or through a voltage source) to find cut points. Such simple schemes, basically producing partitions grouping channel-connected MOSFET devices, proved extremely effective and efficient for predominantly digital circuits connected to simple power supply consisting of ideal voltage sources.

Yet, as focus of fast-SPICE has shifted from large, pre-layout mostly digital circuits with simple power supplies, to mixed-signal, full-chip simulation which has often included post-layout data, efficient partitioning became much more challenging. For circuits with nonideal power supply networks, including internal power supplies (e.g. for phase-locked loops with charge pumps), or circuits with switched power supply, simple partitioning schemes often produced unacceptably large partitions, yielding inadequate simulation performance. In order to work at all older fast-SPICE tools frequently required user-provided information on nodes which one should partition across, as well as performed synchronous solves between different partitions.

Recently much more advanced partitioning techniques have been proposed, providing efficient, automatic partitioning for large-scale complex mixed-signal designs. Technologies, such as the one presented in [3], apply a much more comprehensive view at the system partitioning problem. In this approach elaborate automatic identification of circuit blocks is first performed, e.g. identification of power supply networks (including internal power supplies), as well as of digital circuit regions. Also the graph-based view is used to perform topological analysis of the circuit in order to identify devices which can (or cannot) be partitioned. Finally, matrix view of the problem is applied. In this view, matrix fill-in patterns are automatically reviewed as well as solve/factorization cost analysis is performed in order to find additional partitioning nodes [3]. Again, this comprehensive approach, combining different views at the computational problem, allows one to effectively partition even strongly coupled blocks, and subsequently simulate them in asynchronous manner. Consequently, the discussed technology outperforms many previously developed schemes, particularly for diverse advanced technology circuits, which include low-power designs, internal power supplies, and flash memory arrays.

4.3 Memory Simulation Acceleration

Another area in which fast-SPICE tools are expected to excel is simulation of memory circuits, including arrays ranging from hundreds of thousands to billions of cells (including DRAM, SRAM, FLASH, CAM etc.). Since memory circuits are encountered in virtually any chip, efficient memory simulation is required in order to achieve acceptable performance in full-chip simulations. This in turn implies application of specialized acceleration techniques which can handle memory circuits of enormous size by optimizing the associated simulation scheme and, most importantly, by reducing memory storage requirements.

On one hand, memory circuits are very challenging to simulate, due to a very large size of the corresponding system. Also, for a typical two-dimensional array topology (cf. Fig. 2) the bandwidth of the corresponding system matrix is $O(\sqrt{N})$, where N is the number of cells in the array. This implies that the cost of computing a simple matrix-vector product is $O(N^{1.5})$, which can be prohibitively expensive for large N.

On the other hand though, memory circuits are easy to simulate thanks to their very regular topologies, as well as the fact that in most cases, very large portions of the memory array remain latent, since both read and write operations typically activate only a very small fraction of the total number of cells at any given time.

Fast-SPICE technologies exploit both specific topology of memory arrays as well as their operating pattern in order to dramatically accelerate their simulation.

Initially techniques for accelerating memory circuit simulations focused on 'ideal' pre-layout memory arrays, i.e. arrays which do *not* include parasitic RC elements either inside the cells, or along bitlines or wordlines. Older methods often first identified portions of the array which stayed latent throughout the entire simulations, and subsequently removed those portions to lower memory cost and improve performance of the actual simulation (cf. e.g. Hierarchical Array Reduction [7]). Those techniques also frequently required a designer to manually specify the cell subcircuit, as well as heavily relied on efficient hierarchical representation of the memory circuit. Consequently, the methods were most suited for simulations which verified only portions of the entire memory array.

Subsequently, much more advanced technologies were developed, such as Spice Optimized for Arrays [4]. This method also merges various approaches toward the computational problem, yielding highly efficient memory models. In this technique graph-based approach is used to *automatically* detect a wide range of cell and memory topologies. Then, typical operating patterns of memory circuits are exploited to replace the entire array consisting of N cells, with $O(\sqrt{N})$ models, each model representing an entire row or column of cells. Such approach yields critical reduction of memory usage (from $O(N)$ to $O(\sqrt{N})$), and corresponding performance improvements. Since performance of this optimization technology is also independent of

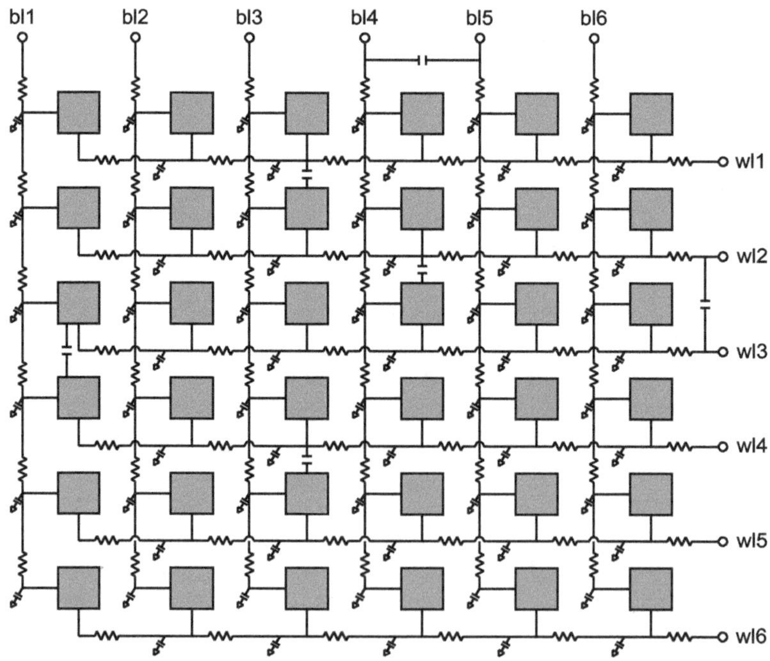

Fig. 2 Model post-layout memory array topology. Grayed boxes represent memory cells. Cells are connected with each other through bitlines and wordlines which are distributed RC lines. Also capacitive couplings are present between different cells, bitlines and wordlines

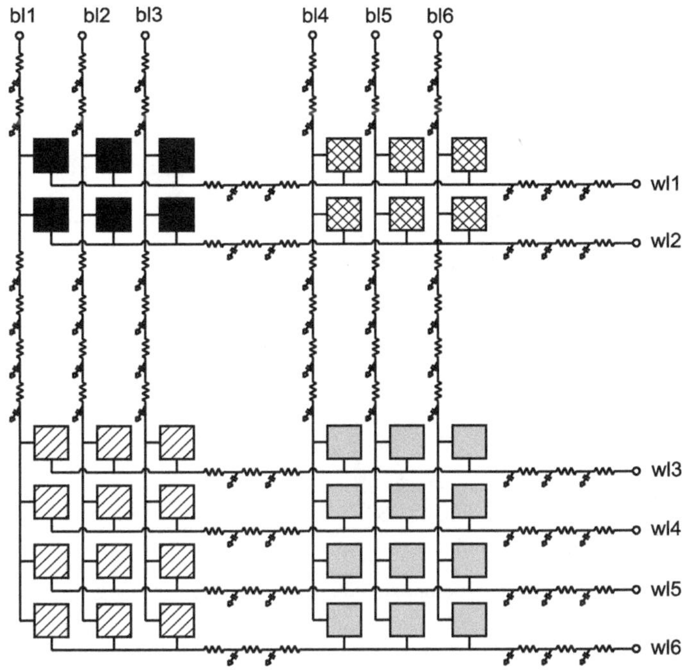

Fig. 3 Cells are automatically reconnected to different nodes located along distributed RC bitlines and wordlines according to information on signal delays along the lines. Consequently, groups of cells which connect to the same bitline or wordline nodes are formed (marked with different patterns). Such groups or subarrays can be efficiently modeled using techniques for pre-layout memory arrays such as Spice Optimized for Arrays [4]

the simulated memory access pattern, it makes the method well suited for extensive memory verification runs [4], which simulate accessing e.g. 100% of cells.

Verification of *post-layout* memory arrays has brought simulation challenges to yet another level. The new issues included:

- Huge number of RC elements inside the array, modeling delays along bitlines and wordlines, and parasitic effects inside the cells;
- Numerous capacitive couplings between bitlines, wordlines and different memory cells (cf. Fig. 2);
- Irregularities in topology of the memory schematic due to varying distribution of parasitic RC elements across different array sub-blocks and cells;
- Varying cell transistor parameters across different cells.

Consequently, simple acceleration techniques based on isomorphism, hierarchical simulation, and idealized array topology for pre-layout memories could not be readily applied in the post-layout case.

Effective technology for accelerated simulation of post-layout arrays required combining numerous state-of-the-art optimization methods, as well as—again—applying diverse views at the computational problem. One such technique, has been

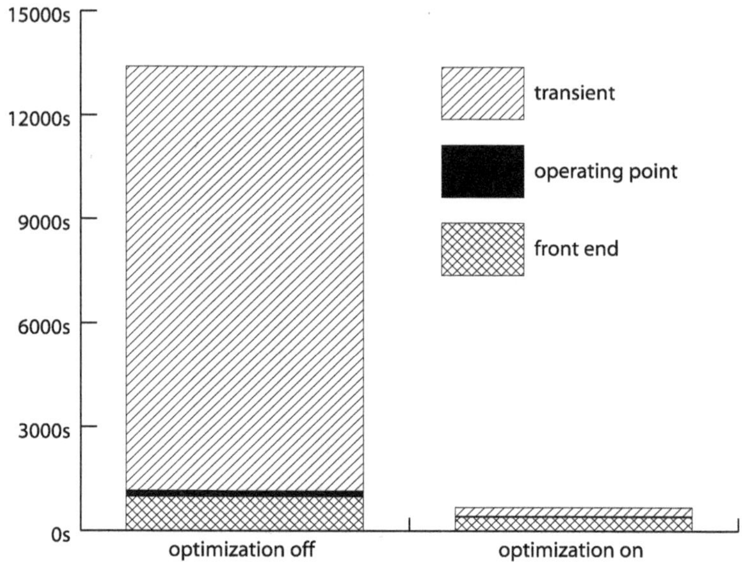

Fig. 4 Reduction of runtime for different simulation phases thanks to post-layout memory optimization technology [9]

presented in [9]. Firstly, it extends automatic graph-based cell and array topological detection to post-layout case. Then, in order to make use of techniques for 'ideal' memory circuits such as [4], it reconnects cells to different nodes located on distributed RC bitlines and wordlines, and hence ideal sub-arrays are formed (cf. Fig. 3). Each of these sub-arrays can be effectively optimized using e.g. method [4]. In order to reconnect cells along bitlines and wordlines while maintaining desired accuracy target, rigorous yet compact models for bitlines and wordlines are automatically generated using matrix-based RC optimization (MOR) algorithms. Fast approximate simulation of read and write operations using those compact bitline and wordline models yields information on feasible cell grouping. Furthermore, automatic optimizations of capacitive couplings between bitlines, wordlines and cells, as well as optimizations of parasitics located inside cells are applied [9].

The comprehensive optimization approach discussed above results in substantial performance gains for simulations verifying post-layout memory circuits. Figure 4 shows runtime reduction when optimization [9] is applied during simulation of a memory circuit with 0.5 million SRAM cells, and 2.6 million parasitic elements (created during post-layout RC extraction performed for the entire memory array). Runtime is reduced at each simulation stage: the setup phase ('front end'), computation of the operating point, and transient simulation, yielding an overall 50X speedup. No less significant are savings in memory required to complete the simulation, as shown in Fig. 5.

Additional advantages of newer memory optimization technologies ([4, 9]) over older approaches become even more apparent when large portions of memory array

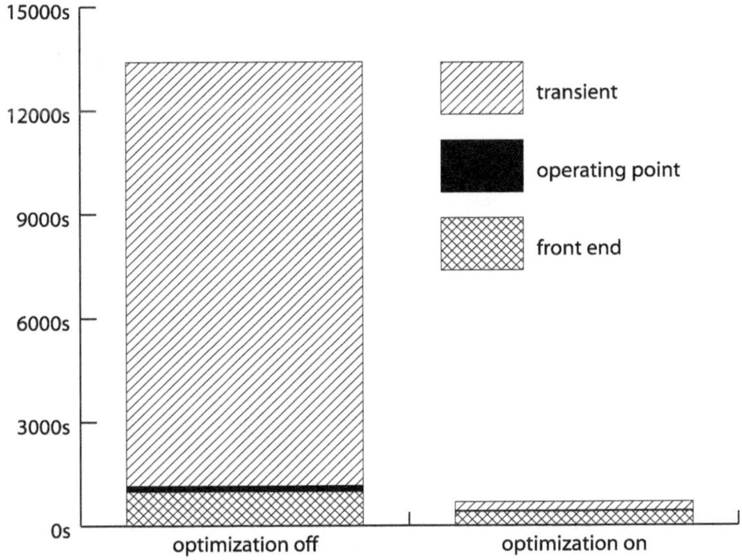

Fig. 5 Reduction of memory usage for different simulation phases thanks to post-layout memory optimization technology [9]

are tested, i.e. when the verification pattern involves accessing (reading/writing) a large percentage of the total number of cells. Figure 6 compares runtimes for a series of simulations with varying percentage of the total number of accessed cells, for an older acceleration technology and the new one [9]. The tests were performed for the same memory circuit as the one described above. Analogous comparisons for memory usage are presented in Fig. 7. First of all, the new technology yields approximately 17-fold speedup over the older solution, for an input pattern which exercises all the cells. Even more importantly, it is apparent from Fig. 7, that while for the older solution the simulator memory usage increases with the increasing percentage of exercised memory cells, it stays constant for the new technology. Hence, the new solution is particularly well suited for simulations which perform extensive memory verification.

5 Challenges of Fast-SPICE and Future Research

Previous sections discussed how different approaches toward the problem of computational modeling of an integrated circuit at transistor level lead to various simulation acceleration technologies embedded in fast-SPICE tools. It has also been shown that combining those approaches or viewpoints toward the simulated system leads to very effective and efficient techniques, which address some of the key problems

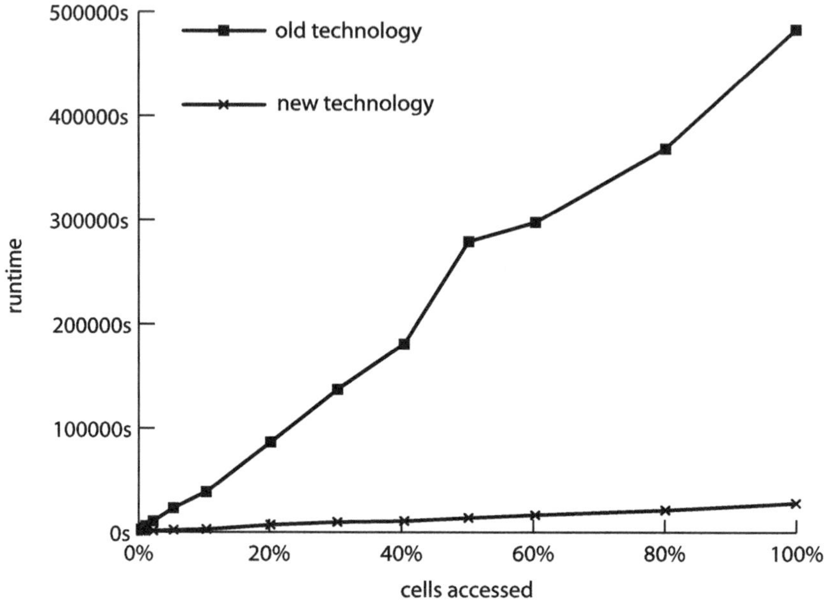

Fig. 6 Comparison of runtimes between simulations applying different memory acceleration techniques—an old one, and a new one [9]. Runtimes are given for a series of simulations with different percentage of the total number of cells accessed

including simulation of large parasitic networks and memory circuits, and effective partitioning of the computational problem.

Despite enormous progress of technology for accelerated transistor-level simulation, new problems keep emerging at an astonishing rate, continually challenging developers and researchers in the fast-SPICE area. Firstly, new challenges arise due to advancing CMOS process technology which makes electrical simulation increasingly more complex. In particular:

- Complexity of MOSFET models keeps increasing, along with variability in transistor model parameters;
- Simulated circuits become increasingly parasitic-dominated, and include massive capacitive couplings which challenge many traditional fast-SPICE heuristics and acceleration methods;
- Ever growing portions of the simulated circuit exhibit analog behavior which requires more careful modeling;
- Power supply networks become increasingly more extensive and complex;
- Architectures for memory circuits become more complicated;
- Element count in full-chip simulations keeps increasing.

Apart from the factors mentioned above, which make the system of equations to be solved larger and more complex, progress in manufacturing technology and IC design also change simulation needs of IC designers. There is a growing demand for simulation capabilities which go beyond traditional fast-SPICE scope, including:

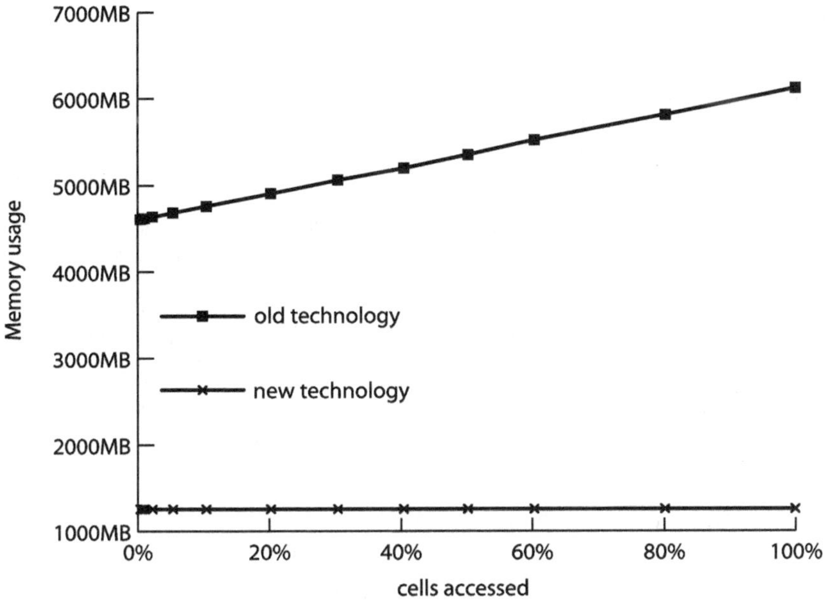

Fig. 7 Comparison of memory usage between simulations applying different memory acceleration techniques—an old one, and a new one [9]. Memory footprints are given for a series of simulations with different percentage of the total number of cells accessed

- Statistical fast-SPICE circuit simulations;
- Chip power-up simulations;
- Reliability analysis, including simulations for current density, electromigration, IR drop, device aging;
- Near-SPICE accuracy in full-chip simulations.

In order to satisfy more sophisticated fast-SPICE simulation needs, as well as growing costs of computations associated with more complicated and larger systems, substantial research efforts are expected on a number of topics. Firstly, more attempts will be made to bridge accuracy gap between SPICE and fast-SPICE tools. Next, tradeoffs associated with new hardware architectures (GPUs, multi-core architectures) will be further investigated. Another focus area will concern techniques for efficient simulation of power supply networks, as well as power-up effects. Finally, fast-SPICE developers are expected to include even more automation in the simulators, in order to improve their ease of use and robustness of the results. As a result of those efforts, a new generation of fast-SPICE tools will undoubtedly emerge, once again pushing performance and capacity limits, and challenging many of today's simulation paradigms.

Acknowledgements The author would like to express his gratitude to Kevin Kerns and Mayukh Bhattacharya for their mentoring throughout the years and for all the valuable discussions, which provided a lot of inspiration for writing this text. Lastly, the author would like to thank Sylwia Rewieńska for her help with preparing this chapter's drawings.

References

1. Cadence White Paper (2004), Using Hierarchy and Isomorphism to Accelerate Circuit Simulation, available at http://w2.cadence.com/whitepapers/5084_AccelCircuitWP_FNL.pdf
2. Celik M, Pileggi L, Odabasioglu A (2002), IC Interconnect Analysis, Kluwer Academic Publishers, Boston
3. Kerns KJ, Bhattacharya M, Rudnaya S, Gullapalli K (2007), Automatic, Hierarchy-Independent Partitioning Method for Transistor-Level Circuit Simulation, patent application submitted to U.S. Patent and Trademark Office (pending)
4. Kerns KJ, Peng Z (2008), SPICE optimized for arrays, U.S. patent no. 7,324,363
5. Karypis G, Kumar V, hMETIS 1.5: A hypergraph partitioning package, Technical Report, Department of Computer Science, Univ. of Minnesota, available at http://www.cs.umn.edu/~metis
6. Kerns KJ, Yang AT (1998), Preservation of passivity during RLC network reduction via split congruence transformations, IEEE Trans. Comput.-Aided Design Integr. Circuits Syst., vol. 17, no. 7, pp. 582–591
7. Synopsys NanoSim User's and Reference Guide, 2004
8. Peng Li, Pileggi L (2002), A linear-centric modeling approach to harmonic balance analysis, in Proc. Design, Automation and Test in Europe Conf., pp. 634–639
9. Rewienski M, Kerns KJ (2007), Optimization of Post-Layout Arrays of Cells for Accelerated Transistor Level Simulation, patent application submitted to U.S. Patent and Trademark Office (pending)
10. Phillips JR, Silveira LM (2005), Poor Man's TBR: A simple model reduction scheme, IEEE Trans. Comput.-Aided Design Integr. Circuits Syst., vol. 24, no. 1, pp. 43–55
11. Rommes J, Schilders WHA (2010), Efficient Methods for Large Resistor Networks, IEEE Trans. Comput.-Aided Design Integr. Circuits Syst., vol. 29, no. 1, pp. 28–39
12. Saad Y (2003), Iterative Methods for Sparse Linear Systems, 2nd ed., SIAM, Philadelphia
13. Sheehan BN (1999), TICER: Realizable reduction of extracted RC circuits, in Proc. IEEE/ACM Int. Conf. Comput.-Aided Design, pp. 200–203
14. Sheehan BN (2007), Realizable Reduction of RC Networks, IEEE Trans. Comput.-Aided Design Integr. Circuits Syst., vol. 26, no. 8, pp. 1393–1407
15. Tcherniaev A, et al. (2003), Transistor level circuit simulator using hierarchical data, U.S. patent no. 6,577,992
16. Zhao Li, Shi C-JR (2006), SILCA: SPICE-accurate iterative linear-centric analysis for efficient time-domain simulation of VLSI circuits with strong parasitic couplings, IEEE Trans. Comput.-Aided Design Integr. Circuits Syst., vol. 25, no. 6, pp. 1087–1103

Recent Advances in Structure-Preserving Model Order Reduction

Roland W. Freund

Abstract In recent years, model order reduction techniques based on Krylov subspaces have become the methods of choice for generating small-scale macromodels of the large-scale multi-port RCL networks that arise in VLSI interconnect analysis. A difficult and not yet completely resolved issue is how to ensure that the resulting macromodels preserve all the relevant structures of the original large-scale RCL networks. In this paper, we present a brief review of how Krylov subspace techniques emerged as the algorithms of choice in VLSI circuit simulation, describe the current state-of-art of structure-preserving model order reduction, discuss some recent advances, and mention open problems.

1 Introduction

In 1990, *Asymptotic Waveform Evaluation* (AWE) was introduced as a generalization of the classical Elmore delay to obtain more accurate timing estimates for the increasingly larger RCL networks arising in VLSI interconnect modeling. It was quickly realized that the mathematical concept behind AWE is equivalent to model order reduction based on Padé approximation. The connection between Krylov subspaces and Padé approximation then provided the basis for devising Krylov subspace-based methods for generating macromodels of large-scale multi-port RCL networks. These algorithms satisfy some of the same key properties as AWE, such as moment matching, but they avoid AWE's unstable numerical behavior due to round-off errors and explicit moment computations.

Generating small-scale macromodels of large-scale RCL networks is a special instance of the problem of *Model Order Reduction* (MOR) of large-scale systems, which is a classical and well-studied subject in the area of *Control Theory* in math-

Roland W. Freund
Department of Mathematics, University of California at Davis, One Shields Avenue, Davis, CA 95616, USA, e-mail: freund@math.ucdavis.edu

P. Li et al. (eds.), *Simulation and Verification of Electronic and Biological Systems*,
DOI 10.1007/978-94-007-0149-6_3,

ematics. However, in control theory, the systems to be reduced are of relatively small or at most moderate size, and the methods developed in this area typically are computationally too expensive to be viable for large-scale systems. The need to efficiently reduce the ever-increasing sizes of the RCL networks used to model VLSI interconnect led to a renewed interest in MOR and especially in the development of algorithms that are applicable to large-scale systems.

Generally, Krylov subspace-based algorithms have emerged as powerful and versatile tools for MOR large-scale systems. Even though many of these algorithms were originally proposed in the context of VLSI interconnect analysis, they have found applications in many other areas, such as structural analysis and computational acoustics. On the other hand, the problem of MOR of large-scale RCL networks has brought the issue of structure preservation to prominence. The ideal MOR algorithm for large-scale RCL networks would generate small-scale macromodels in the form of RCL networks. Krylov subspace-based MOR algorithms, however, are linear algebra techniques, and they produce macromodels described by data matrices, rather than networks. A more modest goal then is to generate macromodels that preserve essential properties of RCL networks, such as passivity and reciprocity.

The *Passive Reduced-order Interconnect Macromodeling Algorithm* (PRIMA) was the first Krylov subspace-based method that produces passive macromodels of RCL networks. The key insight behind PRIMA is that passivity can be preserved by employing explicit projection of the data matrices describing the original RCL network onto Krylov subspaces and by giving up about half the moment matching satisfied by MOR based on Padé approximation. On the other hand, PRIMA does not preserve other structures inherent to RCL networks, such as reciprocity, which makes it harder to synthesize the PRIMA macromodels as actual electrical networks.

The *Structure-Preserving Reduced-order Interconnect Macromodeling* (SPRIM) algorithm was introduced as a structure-preserving variant of PRIMA that overcomes many of the shortcomings of PRIMA and at the same time, is more accurate than PRIMA. The purpose of this paper is twofold. First, we review PRIMA and SPRIM and their basic properties. Second, we discuss some recent advances in structure-preserving MOR. In particular, we describe a thick-restart Krylov subspace approach that allows the use of complex and multiple expansion points for the underlying moment-matching property, and we discuss some techniques designed for the efficient implementation of this approach.

This paper is organized as follows. In Sect. 2, we summarize the standard description of general RCL networks and some of their key properties. In Sect. 3, we present a brief review of the evolution of Krylov subspace-based model order reduction from moment matching to the currently dominating paradigm of projection onto Krylov subspaces. In Sects. 4 and 5, we describe PRIMA and SPRIM, respectively. In Sect. 5, we discuss some recent advances in adapting thick-restart Krylov subspace techniques to the problem of model order reduction. Section 6 addresses some issues arising in moment matching with complex expansion points. Finally, we make some concluding remarks in Sect. 7.

Throughout this paper the following notation is used. The set of real and complex numbers is denoted by $\mathbb{R}$ and $\mathbb{C}$, respectively. For general (real or complex) matrices

$M = [m_{jk}]$, $M^T := [m_{kj}]$ is the *transpose* of M, and $M^H := [\overline{m_{kj}}]$ is the *Hermitian* (or *complex conjugate*) of M. The $n \times n$ identity matrix is denoted by I_n, or simply I if its size is apparent from the context. The zero matrix is denoted by 0; its actual size will always be apparent from the context. The notation $M \succeq 0$ ($M \succ 0$) is used to indicate that a real or complex square matrix M is *Hermitian positive semidefinite* (*positive definite*). If all entries of the matrix $M \succeq 0$ ($M \succ 0$) are real, then M is said to be *symmetric positive semidefinite* (*positive definite*). Finally, for a matrix $V \in \mathbb{R}^{N \times n}$, we denote by colspan V the linear subspace of $\mathbb{R}^N$ spanned by the columns of V. Note that colspan V has dimension n if, and only if, rank $V = n$.

2 Description of RCL Networks

An *RCL network* is an electronic circuit with linear resistors, capacitors, inductors, and independent voltage and current sources as its only elements. In this section, we briefly review the standard description and some key properties of such RCL networks.

2.1 RCL Network Equations

A system of equations describing any electronic circuit can be obtained by combining *Kirchhoff's current laws* (KCLs), *Kirchhoff's voltage laws* (KVLs), and the *branch constitutive relations* (BCRs). The BCRs are the equations that describe the electronic behavior of each of the circuit elements. For example, the BCR of a resistor is Ohm's law.

An elegant approach to formulating KCLs and KVLs for a given electronic circuit is based on the representation of the circuit topology as a directed graph; see, e.g., [9, 32, 33, 35]. The edges of such a graph correspond to the elements of the circuit, and the nodes correspond to the nodes of the circuit, which represent the interconnections of the circuit elements. Each edge e can be expressed as an ordered pair $e = (n_1, n_2)$ of nodes n_1 and n_2, where the direction of e is from n_1 to n_2. For circuit elements for which the direction of the electric current through the element is known beforehand, the direction of the associated edge is chosen accordingly. For any other circuit element, an arbitrary direction is assigned to the associated edge. If the computed electric current through an element is nonnegative, then the current flow is in the direction of the edge; otherwise, the actual current flow is against the direction of the edge. Such a directed graph can be described by its *incidence matrix* the rows and columns of which correspond to the nodes and edges, respectively, and the entries of which are defined as follows. The column associated with edge $e = (n_1, n_2)$ contains the entry "1" in the row position corresponding to node n_1, the entry "-1" in the row position corresponding to node n_2, and zero entries otherwise. Since the resulting matrix is always rank deficient, one deletes the row associated with the *ground node* of the circuit. We refer to the matrix obtained after this deletion as the incidence matrix $\mathscr{A}$ of the circuit. In terms of $\mathscr{A}$, all of the circuit's KCLs and KVLs can be stated compactly as follows:

$$\mathscr{A}\, i_{\mathscr{E}} = 0 \quad \text{and} \quad \mathscr{A}^T v = v_{\mathscr{E}}. \tag{1}$$

Here, $i_{\mathscr{E}}$ is the vector the entries of which are the currents through all the circuit elements, $v_{\mathscr{E}}$ is the vector the entries of which are the voltages across all the circuit elements, and v is the vector the entries of which are the voltages at the nodes of the circuit, except for the ground node at which the voltage is zero.

We now assume that the given electronic circuit is an RCL network. We use subscripts r, c, l, v, and i, to refer to edge quantities corresponding to the resistors, capacitors, inductors, voltage sources, and current sources, respectively, of the given RCL network. Moreover, we assume that the edges of the associated directed graph are ordered according to element type. Consequently, the graph's incidence matrix $\mathscr{A}$ and the vectors $i_{\mathscr{E}}$ and $v_{\mathscr{E}}$ can be partitioned as follows:

$$\mathscr{A} = \begin{bmatrix} \mathscr{A}_r & \mathscr{A}_c & \mathscr{A}_l & \mathscr{A}_v & \mathscr{A}_i \end{bmatrix}, \qquad i_{\mathscr{E}} = \begin{bmatrix} i_r \\ i_c \\ i_l \\ i_v \\ i_i \end{bmatrix}, \qquad v_{\mathscr{E}} = \begin{bmatrix} v_r \\ v_c \\ v_l \\ v_v \\ v_i \end{bmatrix}.$$

Furthermore, the BCRs for the resistors, capacitors, and inductors can be stated in the following compact form:

$$v_r(t) = R\, i_r(t), \qquad i_c(t) = C \frac{d}{dt} v_c(t), \qquad v_l(t) = L \frac{d}{dt} i_l(t). \tag{2}$$

Here, R and C are diagonal matrices, the diagonal entries of which are the resistances of the resistors and the capacitances of the capacitors, respectively, and in particular, $R \succ 0$ and $C \succ 0$. The matrix L contains the inductances between the inductors as its entries. If mutual inductances are included, then L is a full matrix; otherwise, L is also a diagonal matrix. In both cases, $L \succ 0$. Therefore, the matrices in (2) always satisfy

$$R \succ 0, \qquad C \succ 0, \quad \text{and} \quad L \succ 0. \tag{3}$$

Finally, the BCRs for the independent voltage and current sources of the RCL network simply state that

$$v_v(t) \quad \text{and} \quad i_i(t) \tag{4}$$

are given functions that can be chosen as inputs to the network, whereas

$$v_i(t) \quad \text{and} \quad i_v(t) \tag{5}$$

are unknown output functions that need to be determined as part of the problem of solving the system of equations describing the given RCL network.

The KCLs and KVLs (1) together with the BCRs (2) yield a system of equations for the unknown circuit quantities $v(t)$, $v_{\mathscr{E}}$, and $i_{\mathscr{E}}$. The size of this system can be reduced significantly by using some of the relations in (1) and (2) to eliminate v_r, v_c, v_l, i_r, and i_c. This process is known as modified nodal analysis and results in the system of equations

$$\begin{aligned} \mathscr{A}_c C \mathscr{A}_c^T \frac{d}{dt} v(t) + \mathscr{A}_r R^{-1} \mathscr{A}_r^T v(t) + \mathscr{A}_l i_l(t) + \mathscr{A}_v i_v(t) &= -\mathscr{A}_i i_i(t), \\ L \frac{d}{dt} i_l(t) - \mathscr{A}_l^T v(t) &= 0, \\ \mathscr{A}_v^T v(t) &= v_v(t) \end{aligned} \tag{6}$$

for the remaining unknown quantities $v(t)$, $i_l(t)$, and $i_v(t)$; see, e.g., [6, Sect. 5.2] or [21]. Recall from (4) that the functions $i_i(t)$ and $v_v(t)$ appearing on the right-hand side of (6) are given. Furthermore, the unknown output function $v_i(t)$ in (5) can be easily obtained from $v(t)$ via the relation

$$v_i(t) = \mathscr{A}_i^T v(t), \tag{7}$$

which follows from the KVLs (1).

For the purpose of model order reduction, it is convenient to state the equations (6) and (7) as a *descriptor system* of the form

$$\begin{aligned} E \frac{d}{dt} x(t) &= A x(t) + B u(t), \\ y(t) &= B^T x(t). \end{aligned} \tag{8}$$

Here, we have set

$$u(t) := \begin{bmatrix} -i_i(t) \\ v_v(t) \end{bmatrix}, \qquad y(t) := \begin{bmatrix} v_i(t) \\ -i_v(t) \end{bmatrix}, \qquad x(t) := \begin{bmatrix} v(t) \\ i_l(t) \\ i_v(t) \end{bmatrix}, \tag{9}$$

and

$$A := \begin{bmatrix} A_{11} & -\mathscr{A}_l & -\mathscr{A}_v \\ \mathscr{A}_l^T & 0 & 0 \\ \mathscr{A}_v^T & 0 & 0 \end{bmatrix}, \qquad E := \begin{bmatrix} E_{11} & 0 & 0 \\ 0 & L & 0 \\ 0 & 0 & 0 \end{bmatrix}, \qquad B := \begin{bmatrix} \mathscr{A}_i & 0 \\ 0 & 0 \\ 0 & -I \end{bmatrix}, \tag{10}$$

where

$$A_{11} := -\mathscr{A}_r R^{-1} \mathscr{A}_r^T \quad \text{and} \quad E_{11} := \mathscr{A}_c C \mathscr{A}_c^T. \tag{11}$$

The vector-valued functions $u(t)$, $y(t)$, and $x(t)$ are called *input vector*, *output vector*, and *state-space vector* of the descriptor system (8). We denote by m the length of the input and output vectors and by N the length of the state-space vector. The integer N is called the *state-space dimension* of (8). In view of (9), m is the total number of voltage and current sources of the given RCL network, and N is the sum of the number of nodes (excluding the ground node), the number of inductors, and the number of voltage sources. Note that $A,\ E \in \mathbb{R}^{N\times N}$, $B \in \mathbb{R}^{N\times m}$, and the matrix E is singular in general. Finally, we remark that A and E satisfy the semidefiniteness conditions

$$A + A^T = \begin{bmatrix} 2A_{11} & 0 & 0 \\ 0 & 0 & 0 \\ 0 & 0 & 0 \end{bmatrix} \preceq 0 \quad \text{and} \quad E = E^T \succeq 0, \tag{12}$$

which readily follow from (3), (10), and (11).

2.2 RCL Transfer Functions

The first relation of the descriptor system (8) represents a linear system of differential-algebraic equations. In Laplace domain, this relation is transformed into a purely algebraic linear system that can be used to formally eliminate the state-space vector. To this end, we denote by $\hat{x}(s)$, $\hat{u}(s)$, and $\hat{y}(s)$ the Laplace transform of $x(t)$, $u(t)$, and $y(t)$, respectively. Application of the Laplace transform to the time-domain system (8) results in the Laplace-domain system

$$\begin{aligned} sE\hat{x}(s) &= A\hat{x}(s) + B\hat{u}(s), \\ \hat{y}(s) &= B^T\hat{x}(s), \end{aligned} \tag{13}$$

where $s \in \mathbb{C}$.

From now on, we assume that the *matrix pencil* $sE - A$ is *regular*, i.e., the matrix $sE - A$ is singular only for finitely many values of $s \in \mathbb{C}$. We can then eliminate $\hat{x}(s)$ from (13) to obtain the input-output relation

$$\hat{y}(s) = H(s)\,\hat{u}(s), \quad \text{where } H(s) := B^T(sE - A)^{-1}B. \tag{14}$$

The function H is called the *transfer function* of the descriptor system (8). We remark that H is a rational $m \times m$-matrix-valued function and that the potential poles of H are the values of $s \in \mathbb{C}$ for which the matrix $sE - A$ is singular. For RCL networks with only current sources, $\hat{u}(s)$ and $\hat{y}(s)$ are the input currents and output voltages of these sources, and thus $H(s)$ is the *impedance matrix* of the network. Similarly, $H(s)$ is the *admittance matrix* for RCL networks with only voltage sources. For RCL networks with both voltage and current sources, $H(s)$ is called a *hybrid matrix* of the network; see, e.g., [1, Sect. 2.4].

Finally, we remark that the regularity assumption for the matrix pencil $sE - A$ is satisfied for any realistic RCL network. Indeed, $sE - A$ is regular if, and only if, the subnetwork consisting of only the voltage sources of the RCL network has no closed (undirected) loops and the (undirected) graph corresponding to the subnetwork obtained from the RCL network by deleting all its current sources is connected; see, e.g., [20, Theorem 1].

2.3 Passivity

A physical system is said to be *passive* if it only consumes energy. Clearly, any RCL network is passive.

It is well known [1] that passivity of a descriptor system (8) is equivalent to positive realness of its associated transfer function (14), H. Here, a transfer function H is said to be *positive real* if it has no poles in the right half

$$\mathbb{C}_+ := \{ s \in \mathbb{C} \,|\, \operatorname{Re} s > 0 \}$$

of the complex plane and

$$H(s) + \big(H(s)\big)^H \succeq 0 \quad \text{for all } s \in \mathbb{C}_+.$$

Since RCL networks are passive, the associated transfer function H in (14) is positive real. This fact can also be deduced directly from the semidefiniteness conditions (12) by employing the following result, which can be found as Theorem 13 in [16].

Theorem 1. *Let $A, E \in \mathbb{R}^{N \times N}$ and $B \in \mathbb{R}^{N \times m}$ be given matrices, and assume that $A + A^T \preceq 0$, $E = E^T \succeq 0$, and that the matrix pencil $sE - A$ is regular. Then, the function*

$$H(s) := B^T (sE - A)^{-1} B$$

is positive real.

2.4 Reciprocity

A second important property of RCL networks is *reciprocity*, which represents a certain symmetry of the input-output behavior of the network; see, e.g., [36] and [1, Sect. 2.8]. For RCL networks with only current sources, reciprocity is equivalent to the symmetry of the impedance matrix $H(s)$, i.e., $H(s) = H(s)^T$ for all $s \in \mathbb{C}$. Similarly, reciprocity is equivalent to the symmetry of the admittance matrix $H(s)$ for RCL networks with only voltage sources.

For general RCL networks with m_i current and m_v voltages sources, reciprocity is equivalent to the transfer function $H(s)$ being symmetric with respect to the *signature matrix*

$$\Sigma := \begin{bmatrix} I_{m_i} & 0 \\ 0 & -I_{m_v} \end{bmatrix} \in \mathbb{R}^{m \times m}, \qquad m = m_i + m_v. \tag{15}$$

Here, H is said to be *symmetric with respect to* Σ if $H(s)\Sigma = \Sigma(H(s))^T$ for all $s \in \mathbb{C}$. Note that the definition (15) of the signature matrix assumes the ordering (9) of the input and output vectors u and y. If a different ordering for u and y is used, then the diagonal entries of Σ need to be permuted accordingly.

It turns out that reciprocity follows easily from the block structure of the matrices A, E, and B defined in (10) and from the symmetry of the diagonal blocks of A and E. In fact, we have the following theorem the proof of which is straightforward and left as an exercise to the reader.

Theorem 2. *Let* $A, E \in \mathbb{R}^{N\times N}$ *and* $B \in \mathbb{R}^{N\times m}$ *be matrices with block structures of the form*

$$A = \begin{bmatrix} A_{11} & -A_{12} & -A_{13} \\ A_{12}^T & 0 & 0 \\ A_{13}^T & 0 & 0 \end{bmatrix}, \qquad E = \begin{bmatrix} E_{11} & 0 & 0 \\ 0 & E_{22} & 0 \\ 0 & 0 & 0 \end{bmatrix}, \qquad B = \begin{bmatrix} B_{11} & 0 \\ 0 & 0 \\ 0 & B_{32} \end{bmatrix},$$

where the sizes of the blocks are the same as in (10) and

$$A_{11} = A_{11}^T, \qquad E_{11} = E_{11}^T, \qquad E_{22} = E_{22}^T.$$

Then, the function

$$H(s) := B^T(sE - A)^{-1}B$$

is symmetric with respect to the signature matrix Σ *defined in (15).*

3 A Brief Review of Krylov Subspace-Based Model Order Reduction

In this section, we present a brief review of the evolution of Krylov subspace-based model order reduction from moment matching to the currently dominating paradigm of projection onto Krylov subspaces.

3.1 Moment Matching and Padé Approximation

The Elmore delay [11] is a classical, simple metric for estimating signal delays in electronic networks. In 1990, *Asymptotic Waveform Evaluation* (AWE) [31] was introduced as a generalization of the Elmore delay to obtain more accurate timing estimates for the large RCL networks arising in VLSI interconnect modeling; see, e.g., [6].

To illustrate the AWE approach, we consider the special case of RC networks driven by a single voltage source. The transfer function H is then scalar-valued and represents the admittance of the RC network. The Taylor expansion of H about $s_0 = 0$ is given by

$$H(s) = \mu_0 + \mu_1 s + \mu_2 s^2 + \cdots + \mu_j s^j + \cdots, \tag{16}$$

where the μ_j's are the so-called *moments*. Suppose we are trying to approximate H by a rational function H_1 of the form

$$H_1(s) = \frac{a_1}{s - b_1}, \quad \text{where } a_1, b_1 \in \mathbb{R}. \tag{17}$$

Given the first two moments, μ_0 and μ_1, of the transfer function (16), H, we can determine values of the parameters a_1, b_1 in (17) such that the Taylor expansions of H_1 and H about $s_0 = 0$ agree in the first two moments, i.e.,

$$H_1(s) = H(s) + \mathcal{O}(s^2). \tag{18}$$

The last condition is satisfied if, and only if, $a_1 = -\mu_0^2/\mu_1$ and $b_1 = \mu_0/\mu_1$. Moreover, one can show that $\mu_0 > 0$ and $\mu_1 < 0$, and thus $a_1 > 0$ and $b_1 < 0$. It follows that

$$R := -\frac{b_1}{a_1} > 0 \quad \text{and} \quad C := \frac{a_1}{b_1^2} > 0.$$

Furthermore, it is easy to verify that for these values of R and C, the approximate transfer function (17), H_1, is the admittance of the simple RC network shown in Fig. 1. The impulse response of this network is given by $i(t) = i(0)\exp(-t/\tau)$, where $\tau := RC$ is the Elmore delay of the original given RC network. Note that the

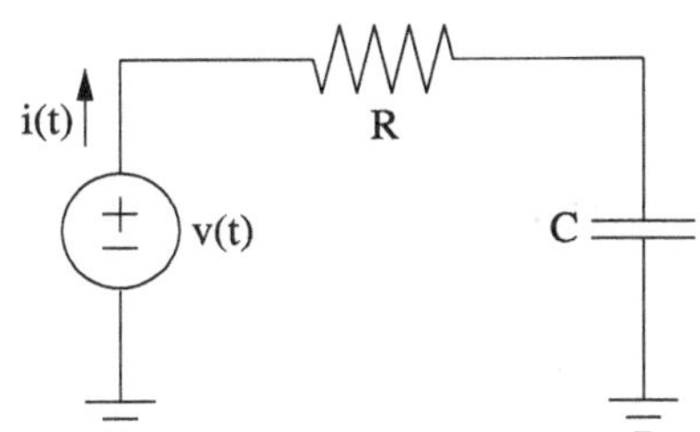

Fig. 1 Synthesis of the approximate transfer function (17), H_1, as an RC network

RC network in Fig. 1 represents a synthesis of the approximation defined by the moment-matching property (18).

AWE generalizes the 'one-pole' approximation (17) of the transfer function (16), H, of the given RC network to a general approximation H_n with n poles. More precisely, H_n is of the form

$$H_n(s) = \frac{a_1}{s - b_1} + \frac{a_2}{s - b_2} + \cdots + \frac{a_n}{s - b_n}, \tag{19}$$

where the $2n$ free parameters $a_1, a_2, \dots, a_n$ and $b_1, b_2, \dots, b_n$ are now chosen such that the Taylor expansions of H_n and H about $s_0 = 0$ agree in the first $2n$ moments, i.e.,

$$H_n(s) = H(s) + \mathcal{O}\big(s^{2n}\big). \tag{20}$$

The property (20) is often referred to as *moment matching*. The corresponding transfer function H_n is an optimal approximation to the original transfer function H in the sense that among all functions of the form (19), H_n matches the maximum possible number of moments. Such optimal approximations are called *Padé approximants* [5].

There are two major issues with the original AWE approach. First, the computation of H_n via explicitly generating the $2n$ moments $\mu_0, \mu_1, \dots, \mu_{2n-1}$ is extremely sensitive to numerical round-off errors and thus is viable only for very small values of n, typically $n \leq 10$; we refer the reader to [12, 13] for a discussion of this issue and for numerical examples. The problem can be resolved by generating the necessary moment information implicitly, by exploiting the connection of moment matching and Krylov subspace methods; see Sect. 3.3.

The second, more fundamental issue is that for general RCL networks, reduced-order transfer functions defined via Padé approximation do not preserve essential properties, such as passivity and reciprocity, of the original network. In fact, preservation of these properties can only be guaranteed for the special cases of RC, RL, and LC networks; see [4, 22, 23]. For general RCL networks, this issue can be resolved by relaxing the moment-matching property and using certain *Padé-type approximants*, instead of Padé approximants; see Sect. 3.5.

3.2 Reduced-Order Models

We now return to the case of general RCL networks with a total number of m independent current and voltage sources. Recall that in time domain, the RCL network equations can be stated as a descriptor system (8) with data matrices $A, E \in \mathbb{R}^{N \times N}$ and $B \in \mathbb{R}^{N \times m}$ of the form (10), where N is the state-space dimension of the system (8). In Laplace domain, the RCL network is described by its transfer function (14), H.

A general *reduced-order model* of (8) is a descriptor system of the same form as (8), but with state-space dimension n $(< N)$, instead of N. Thus a reduced-order model is of the form

$$\begin{aligned} E_n \frac{d}{dt} x_n(t) &= A_n x_n(t) + B_n u(t), \\ y_n(t) &= B_n^T x_n(t), \end{aligned} \tag{21}$$

where A_n, $E_n \in \mathbb{R}^{n \times n}$ and $B_n \in \mathbb{R}^{n \times m}$. Note that the input vector $u(t)$ in (21) is the same as in (8). In particular, the number m is unchanged from (8). The output vector

$y_n(t)$ of (21), however, is only an approximation to the original output vector $y(t)$ of (8). In fact, the problem of model order reduction is to find a sufficiently large reduced state-space dimension n and matrices A_n, E_n, and B_n such that the output vector of the reduced-order model (21) is a 'sufficiently good' approximation to the output vector of the original system (8).

Provided that the matrix pencil

$$sE_n - A_n, \quad s \in \mathbb{C}, \tag{22}$$

associated with the reduced-order model (21) is regular, we have, in analogy to (14), the Laplace-domain input-output relation

$$\hat{y}_n(s) = H_n(s)\,\hat{u}(s), \quad \text{where } H_n(s) := B_n^T(sE_n - A_n)^{-1}B_n, \tag{23}$$

for the reduced-order model (21). Together with (14), it follows that

$$\hat{y}_n(s) - \hat{y}(s) = \big(H_n(s) - H(s)\big)\hat{u}(s).$$

This relation shows that in terms of transfer functions, the problem of model order reduction is to find a sufficiently large reduced state-space dimension n and matrices A_n, E_n, and B_n such that the transfer function (23), H_n, of the reduced-order model is a 'sufficiently good' approximation to the transfer function (14), H, of the original system:

$$H_n(s) \approx H(s) \quad \text{in 'some sense'}.$$

There are many general-purpose methods for constructing suitable approximations H_n; see, e.g., [2, 3]. For the special case of systems describing RCL networks, techniques based on moment matching are the most widely used and in fact, for very large N, are often the only computationally viable approaches.

3.3 Moment Matching Via Krylov Subspace Methods

Recall that for general RCL networks with a total number of m independent current and voltage sources, the transfer functions H and H_n are $m \times m$-matrix-valued. We now consider the extension of the moment-matching property (20) to the case of such general RCL networks. Furthermore, instead of the expansion point $s_0 = 0$ in (20), we allow general expansion points s_0 subject to the constraint that the matrix $s_0 E - A$ is nonsingular. This condition guarantees that s_0 is not a pole of the transfer function (14), H. For now, we restrict ourselves to real expansion points $s_0 \in \mathbb{R}$; the case of general complex expansion points $s_0 \in \mathbb{C}$ is discussed in Sect. 7.

In this general setting, moment matching takes on the following form:

$$H_n(s) = H(s) + \mathscr{O}\big((s - s_0)^q\big). \tag{24}$$

Here, $q = q(n,m)$ is an integer that depends on the reduced state-space dimension n, on m, and on the particular method for constructing the reduced-order model associated with H_n. In this setting, the AWE approach is obtained by choosing a method that gives the maximum value of q in (24). The corresponding function H_n is then called an *n-th Padé approximant* of H and the system (21) a *Padé reduced-order model* of (8). We denote by $\tilde{q}(n.m)$ the maximum value of q in (24). It is well known (see, e.g., [14, 15]) that

$$\tilde{q}(n.m) \geq 2 \left\lfloor \frac{n}{m} \right\rfloor, \tag{25}$$

where equality holds true in the generic case.

As mentioned before, generating Padé reduced-order models directly from the first $\tilde{q} = \tilde{q}(n,m)$ moments μ_j, $j = 0,1,\dots,\tilde{q}-1$, in the expansion

$$H(s) = \mu_0 + \mu_1 (s - s_0) + \mu_2 (s - s_0)^2 + \cdots + \mu_{\tilde{q}-1} (s - s_0)^{\tilde{q}-1} + \cdots \tag{26}$$

is extremely sensitive to numerical round-off errors. The remedy is to avoid the computation of the moments and instead, to generate the information contained in the moments via Krylov subspace techniques. To this end, we rewrite the transfer function (14), H, as follows:

$$\begin{aligned} H(s) &= B^T \left(s_0 E - A + (s - s_0)E\right)^{-1} B = B^T \left(I - (s - s_0)M\right)^{-1} R \\ &= \sum_{j=0}^{\infty} B^T M^j R \, (s - s_0)^j, \end{aligned} \tag{27}$$

where

$$M := -(s_0 E - A)^{-1} E \quad \text{and} \quad R := (s_0 E - A)^{-1} B. \tag{28}$$

Comparing (26) and (27), it follows that

$$\mu_j = B^T M^j R = \left((M^T)^j B\right)^T R, \quad j = 0,1,\dots . \tag{29}$$

The representation (29) shows that the leading coefficients in the expansion (26) of H can be obtained by computing suitable inner products of the leading columns of the matrices

$$\begin{bmatrix} R & MR & M^2 R & \cdots & M^{j-1} R & \cdots \end{bmatrix} \tag{30}$$

and

$$\left[B \;\; M^T B \;\; \left(M^T\right)^2 B \;\; \cdots \;\; \left(M^T\right)^{j-1} B \;\; \cdots \right]. \tag{31}$$

Working directly with the columns of these matrices would result in a procedure that is still sensitive to numerical round-off errors. Instead, Krylov subspace methods generate more suitable basis vectors for the subspaces spanned by the leading columns of these matrices. Computing moment information via inner products of these basis vectors then avoids the numerical issues of moment matching and leads to algorithms that generate Padé reduced-order models in a numerically stable and reliable way; see [12–14, 17, 25] for more details.

For later use, we give a formal definition of the Krylov subspaces associated with the matrices (30) and (31). Let $N_{\max}$ $(\leq N)$ denote the rank of the matrix (30). Then, for $\hat{n} = 1, 2, \ldots, N_{\max}$, the $\hat{n}$-th (right) *Krylov subspace* (induced by M and R) is defined as the $\hat{n}$-dimensional subspace of $\mathbb{R}^N$ spanned by the first $\hat{n}$ linearly independent columns of the matrix (30). In the following, $\mathscr{K}_{\hat{n}}(M,R)$ denotes this $\hat{n}$-th Krylov subspace. Similarly, the $\hat{n}$-th (left) Krylov subspace, $\mathscr{K}_{\hat{n}}(M^T,B)$, is spanned by the first $\hat{n}$ linearly independent columns of the matrix (31).

An important issue in moment matching is the choice of the expansion point s_0 in (24). Typically, the goal is to construct a reduced-order model (21) such that its transfer function H_n approximates the original transfer function H well enough in some given frequency range of interest of the form

$$s = \mathtt{i}\,\omega, \quad \omega_{\min} \leq \omega \leq \omega_{\max},$$

where $0 \leq \omega_{\min} < \omega_{\max}$ are given and $\mathtt{i} = \sqrt{-1}$. Moreover, the state-space dimension n of the reduced-order model should be as small as possible, The general convergence theory of Padé approximation suggests to place s_0 close to the frequency range of interest, yet at a safe distance from the poles of the original transfer function H. Recall from Sect. 2.3 that the poles of H are to the left of or on the imaginary axis. Therefore, the expansion point s_0 is placed in the right half $\mathbb{C}_+$ of the complex plane. The ideal placement of s_0 would be on a horizontal line through the midpoint of the frequency range of interest, resulting in a non-real s_0. This in turn would result in complex matrices M and R in (28) and thus the need for complex rather than real arithmetic, which would increase the computational work by roughly a factor of 4. This is the reason for the usual restriction to positive $s_0 > 0$; see, e.g., the discussion in [13]. A typical placement of $s_0 > 0$ relative to the frequency range of interest is shown in Fig. 2.

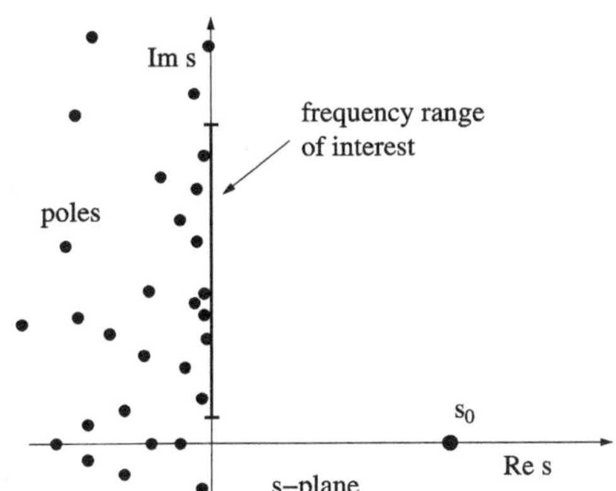

Fig. 2 Typical placement of expansion point s_0 for the moment-matching property (24)

3.4 Passive Models Via Projection

While Padé reduced-order models are optimal in terms of moment matching, they do not preserve the passivity of general RCL networks.

A very basic, general approach to constructing passive reduced-order models is to employ *projection*. Let

$$V_n \in \mathbb{R}^{N\times n} \quad \text{with rank } V_n = n \tag{32}$$

be given. Then, by simply setting

$$A_n := V_n^T A V_n, \qquad E_n := V_n^T E V_n, \quad \text{and} \quad B_n := V_n^T B, \tag{33}$$

one obtains a reduced-order model (21) that can be viewed as a projection of the N-dimensional state space of the original system onto the n-dimensional subspace spanned by the columns of the matrix V_n. In particular, projection employs an ansatz of the form

$$x_n(t) = V_n x(t)$$

for the state-space vector $x_n(t)$ of the reduced-order model (21). Recall from (9) that $x(t)$ denotes the state-space vector of the original descriptor system (8).

Reduced-order models obtained by means of this simple projection approach trivially preserve passivity of the original system; see, e.g., [16, 28–30]. Indeed, the only additional condition on the matrix (32), V_n, is that the resulting matrix pencil (22) is regular. Recall that A and E satisfy the semidefiniteness properties (12). The definitions of A_n and E_n in (33) readily imply that these matrices satisfy the very same semidefiniteness conditions. Therefore, by Theorem 1, the transfer function (23), H_n is positive real, and thus the corresponding reduced-order model (21) is passive.

3.5 Projection Combined with Krylov Subspaces

The projection approach described in Sect. 3.4 can be combined with the use of Krylov subspaces to obtain reduced-order models that are passive and at the same time satisfy a moment-matching property, albeit a weaker one than Padé models.

To this end, we choose the projection matrix (32), V_n, such that

$$\mathcal{K}_{\hat{n}}(M,R) \subseteq \text{colspan } V_n. \tag{34}$$

Recall that $\mathcal{K}_{\hat{n}}(M,R)$ is the $\hat{n}$-th (right) Krylov subspace defined in Sect. 3.3. Since $\mathcal{K}_{\hat{n}}(M,R)$ has dimension $\hat{n}$ and, by (32), V_n is assumed to have rank n, the condition (34) implies that $\hat{n} \leq n$. Note that $\hat{n} = n$ if, and only if, equality holds true in (34). In this case, the projection matrix (32), $V_{\hat{n}}$, is a *basis matrix* for $\mathcal{K}_{\hat{n}}(M,R)$, i.e., the columns of $V_{\hat{n}}$ form a basis for the linear subspace $\mathcal{K}_{\hat{n}}(M,R)$. We will use the notation $\hat{V}_{\hat{n}}$, instead of $V_{\hat{n}}$, to indicate that $\hat{V}_{\hat{n}}$ is a basis matrix. We remark that there are standard Krylov subspace techniques, such as the band Arnoldi process [16, Algorithm 1], to construct the columns of $\hat{V}_{\hat{n}}$.

We now return to the general case (34). Let A_n, E_n, and B_n be the matrices defined by the projection approach (33), and let H_n be the transfer function (23) of the corresponding reduced-order model (21). The main result of Krylov subspace-based projection then states that H_n satisfies the moment-matching property

$$H_n(s) = H(s) + \mathcal{O}\left((s - s_0)^{\hat{q}(\hat{n},m)}\right), \quad \text{where } \hat{q}(\hat{n},m) \geq \left\lfloor \frac{\hat{n}}{m} \right\rfloor. \tag{35}$$

Moreover, in the generic case, $\hat{q}(\hat{n},m) = \lfloor \hat{n}/m \rfloor$ in (35). Recall from (24) and (25) that Padé reduced-order models match $\tilde{q}(n,m)$ moments, where

$$\tilde{q}(n,m) \geq 2\left\lfloor \frac{n}{m} \right\rfloor \geq 2\left\lfloor \frac{\hat{n}}{m} \right\rfloor.$$

Thus, Krylov subspace-based projection results in reduced-order models that, in the generic case, match only half as many moments as Padé models if $\hat{n} = n$ and even less if $\hat{n} < n$. Reduced-order transfer functions H_n that match fewer moments than n-th Padé approximants are called *n-th Padé-type approximants* and the system (21) is a *Padé-type reduced-order model* of (8).

The Padé-type moment matching property (35) of Krylov subspace-based projection is well known. For example, it was established for various special cases in [8, 25, 28]. Proofs for the general case can be found in [16, 19].

4 PRIMA

The simplest way to satisfy condition (34) is to choose the projection matrix (32) as a basis matrix for $\mathcal{K}_{\hat{n}}(M,R)$. In this case $\hat{n} = n$, and the moment-matching property (35) takes on the following form:

$$H_{\hat{n}}(s) = H(s) + \mathcal{O}\left((s - s_0)^{\hat{q}(\hat{n},m)}\right), \quad \text{where } \hat{q}(\hat{n},m) \geq \left\lfloor \frac{\hat{n}}{m} \right\rfloor. \tag{36}$$

Moreover, in the generic case, $\hat{q}(\hat{n},m) = \lfloor \hat{n}/m \rfloor$. The corresponding reduced-order models (21) given by the reduced data matrices (33) (where $n = \hat{n}$) are then the ones produced by PRIMA (Passive Reduced-order Interconnect Macromodeling Algorithm) [29, 30].

Next, we present an algorithmic statement of the basic steps of PRIMA.

Algorithm 3 (PRIMA for general RCL networks)

- ***Input:*** *Matrices of the form*

$$A = \begin{bmatrix} A_{11} & -\mathscr{A}_l & -\mathscr{A}_v \\ \mathscr{A}_l^T & 0 & 0 \\ \mathscr{A}_v^T & 0 & 0 \end{bmatrix}, \qquad E = \begin{bmatrix} E_{11} & 0 & 0 \\ 0 & L & 0 \\ 0 & 0 & 0 \end{bmatrix}, \qquad B = \begin{bmatrix} \mathscr{A}_i & 0 \\ 0 & 0 \\ 0 & -I \end{bmatrix},$$

where $A_{11} \preceq 0$, $E_{11} \succeq 0$, *and* $L \succ 0$.
An expansion point $s_0 \in \mathbb{R}$ *such that the matrix* $s_0 E - A$ *is nonsingular; this last condition is satisfied for any* $s_0 > 0$.

- *Formally set*

$$M = -(s_0 E - A)^{-1} E, \qquad R = (s_0 E - A)^{-1} B.$$

- *Until* $\hat{n}$ *is large enough, run your favorite Krylov subspace method (applied to* M *and* R*) to construct a matrix*

$$\hat{V}_{\hat{n}} = \begin{bmatrix} v_1 & v_2 & \cdots & v_{\hat{n}} \end{bmatrix}$$

the columns of which span the $\hat{n}$*-th Krylov subspace* $\mathscr{K}_{\hat{n}}(M,R)$*, i.e.,*

$$\operatorname{colspan} \hat{V}_{\hat{n}} = \mathscr{K}_{\hat{n}}(M,R).$$

- *Set*

$$A_{\hat{n}} = \hat{V}_{\hat{n}}^T A \hat{V}_{\hat{n}}, \qquad E_{\hat{n}} = \hat{V}_{\hat{n}}^T E \hat{V}_{\hat{n}}, \quad \text{and} \quad B_{\hat{n}} = \hat{V}_{\hat{n}}^T B.$$

- ***Output:*** *The data matrices* $A_{\hat{n}}$, $E_{\hat{n}}$, *and* $B_{\hat{n}}$ *of the PRIMA reduced-order model*

$$\begin{aligned} E_{\hat{n}} \frac{d}{dt} x_{\hat{n}}(t) &= A_{\hat{n}} x_{\hat{n}}(t) + B_{\hat{n}} u(t), \\ y_{\hat{n}}(t) &= B_{\hat{n}}^T x_{\hat{n}}(t). \end{aligned} \tag{37}$$

Since the PRIMA reduced-order models (37) are generated via projection onto Krylov subspaces, they satisfy the moment-matching property (36) and they are guaranteed to be passive. In general, however, the PRIMA models do not satisfy reciprocity [34] and thus cannot be synthesized as RCL networks. Furthermore, the reduced-order data matrices $A_{\hat{n}}$, $E_{\hat{n}}$, and $B_{\hat{n}}$ are dense in general and thus do not preserve the block structures of the original data matrices A, E, and B.

5 SPRIM

In this section, we describe the SPRIM (Structure-Preserving Reduced-order Interconnect Macromodeling) algorithm [18, 21], which remedies the lack of structure preservation of PRIMA.

5.1 Preserving Block Structures

As in PRIMA, the main computational step of SPRIM is the generation of a basis matrix $\hat{V}_{\hat{n}}$ for the $\hat{n}$-th block Krylov subspace $\mathscr{K}_{\hat{n}}(M,R)$. In contrast to PRIMA, however, SPRIM does not use the matrix $\hat{V}_{\hat{n}}$ directly for the projection of the original data matrices. Instead, SPRIM employs a modified version of this matrix that trivially leads to structure preservation. To this end, $\hat{V}_{\hat{n}}$ is first partitioned as follows:

$$\hat{V}_{\hat{n}} = \begin{bmatrix} V^{(1)} \\ V^{(2)} \\ V^{(3)} \end{bmatrix}. \tag{38}$$

Here, the block sizes correspond to the block sizes of the original data matrices A and E in (10). While $\hat{V}_{\hat{n}}$ has full column rank $\hat{n}$, the same is not necessarily true for the three subblocks $V^{(l)}$, $l = 1,2,3$, in (38). In particular, the third block, $V^{(3)}$, is of size $m_v \times \hat{n}$, where m_v denotes the number of voltage sources of the given RCL circuit. Usually, the number m_v is very small and $m_v < \hat{n}$. Therefore, $V^{(3)}$ typically does not have full column rank. In the actual implementation of SPRIM, we run a Gram-Schmidt algorithm on the rows of $V^{(3)}$ to determine a matrix $\tilde{V}^{(3)}$ the columns of which span the same space as the columns of $V^{(3)}$, but which has full column rank. The other two blocks usually have many more rows than columns, and these blocks are unlikely not to have full column rank. In the implementation of SPRIM, there is also the option to check the column ranks of the first two blocks and replace them by matrices $\tilde{V}^{(1)}$ and $\tilde{V}^{(2)}$ of full column rank. Next, we set up the actual projection matrix V_n as follows:

$$V_n := \begin{bmatrix} \tilde{V}^{(1)} & 0 & 0 \\ 0 & \tilde{V}^{(2)} & 0 \\ 0 & 0 & \tilde{V}^{(3)} \end{bmatrix}. \tag{39}$$

By construction, we have

$$\mathscr{K}_{\hat{n}}(M,R) = \operatorname{colspan} \hat{V}_{\hat{n}} \subseteq \operatorname{colspan} V_n. \tag{40}$$

Thus the matrix (39), V_n, satisfies condition (34), which in turn guarantees the moment-matching property (35). Furthermore, in view of the block structure of V_n, the data matrices (33) of the resulting reduced-order model obtained via projection with V_n have the same block structure as the original data matrices A, E, and B.

5.2 The Algorithm

An algorithmic statement of the basic steps of SPRIM is as follows.

Algorithm 4 (SPRIM for general RCL networks)

- ***Input:*** *Matrices of the form*

$$A = \begin{bmatrix} A_{11} & -\mathscr{A}_l & -\mathscr{A}_v \\ \mathscr{A}_l^T & 0 & 0 \\ \mathscr{A}_v^T & 0 & 0 \end{bmatrix}, \qquad E = \begin{bmatrix} E_{11} & 0 & 0 \\ 0 & L & 0 \\ 0 & 0 & 0 \end{bmatrix}, \qquad B = \begin{bmatrix} \mathscr{A}_i & 0 \\ 0 & 0 \\ 0 & -I \end{bmatrix},$$

where $A_{11} \preceq 0$, $E_{11} \succeq 0$, *and* $L \succ 0$.
An expansion point $s_0 \in \mathbb{R}$ *such that the matrix* $s_0 E - A$ *is nonsingular; this last condition is satisfied for any* $s_0 > 0$.

- *Formally set*

$$M = -\left(s_0 E - A\right)^{-1} E, \qquad R = \left(s_0 E - A\right)^{-1} B.$$

- *Until* $\hat{n}$ *is large enough, run your favorite Krylov subspace method (applied to* M *and* R*) to construct a matrix*

$$V_{\hat{n}} = \begin{bmatrix} v_1 & v_2 & \cdots & v_{\hat{n}} \end{bmatrix}$$

the columns of which span the $\hat{n}$*-th Krylov subspace* $\mathscr{K}_{\hat{n}}(M,R)$, *i.e.,*

$$\operatorname{colspan} \hat{V}_{\hat{n}} = \mathscr{K}_{\hat{n}}(M,R).$$

- *Let*

$$\hat{V}_{\hat{n}} = \begin{bmatrix} V^{(1)} \\ V^{(2)} \\ V^{(3)} \end{bmatrix}$$

be the partitioning of $\hat{V}_{\hat{n}}$ *corresponding to the block sizes of* A *and* E.

- *For* $l = 1,2,3$ *do:*
If $r_l := \operatorname{rank} V^{(l)} < \hat{n}$, *determine an* $N \times r_l$ *matrix* $\tilde{V}^{(l)}$ *with*

$$\operatorname{colspan} \tilde{V}^{(l)} = \operatorname{colspan} V^{(l)} \quad \textit{and} \quad \operatorname{rank} \tilde{V}^{(l)} = r_l.$$

- *Set*

$$\tilde{A}_{11} = \left(\tilde{V}^{(1)}\right)^T A_{11} \tilde{V}^{(1)}, \qquad \tilde{\mathscr{A}}_l = \left(\tilde{V}^{(1)}\right)^T \mathscr{A}_l \tilde{V}^{(2)}, \qquad \tilde{\mathscr{A}}_v = \left(\tilde{V}^{(1)}\right)^T \mathscr{A}_v \tilde{V}^{(3)},$$

$$\tilde{E}_{11} = \left(\tilde{V}^{(1)}\right)^T E_{11} \tilde{V}^{(1)}, \qquad \tilde{L} = \left(\tilde{V}^{(2)}\right)^T L \tilde{V}^{(2)}, \qquad \tilde{\mathscr{A}}_i = \left(\tilde{V}^{(1)}\right)^T \mathscr{A}_i.$$

- ***Output:*** *The data matrices*

$$A_n = \begin{bmatrix} \tilde{A}_{11} & -\tilde{\mathscr{A}}_l & -\tilde{\mathscr{A}}_v \\ \tilde{\mathscr{A}}_l^T & 0 & 0 \\ \tilde{\mathscr{A}}_v^T & 0 & 0 \end{bmatrix}, \qquad E_n = \begin{bmatrix} \tilde{E}_{11} & 0 & 0 \\ 0 & \tilde{L} & 0 \\ 0 & 0 & 0 \end{bmatrix},$$

$$\textit{and} \quad B_n = \begin{bmatrix} \tilde{\mathscr{A}}_i & 0 \\ 0 & 0 \\ 0 & -(\tilde{V}^{(3)})^T \end{bmatrix}$$

of the SPRIM reduced-order model

$$
\begin{aligned}
E_n \frac{d}{dt} x_n(t) &= A_n x_n(t) + B_n u(t), \\
y_n(t) &= B_n^T x_n(t).
\end{aligned}
\tag{41}
$$

5.3 Some Properties

In this subsection, we list some of the properties of the SPRIM Algorithm 4.

Since the SPRIM reduced-order models (41) are generated via projection onto Krylov subspaces, they are guaranteed to be passive and they satisfy the moment-matching property (35). In contrast to the PRIMA models, the data matrices A_n, E_n, and B_n of the SPRIM reduced-order models exhibit the same block structures as the original data matrices A, E, and B. In particular, the conditions of Theorem 2 hold true for the matrices A_n, E_n, and B_n, and consequently, the SPRIM reduced-order models are guaranteed to satisfy reciprocity. While reciprocity is not sufficient to ensure the synthesizability of reduced-order models as RCL networks, it significantly simplifies the synthesis of the models as simple electronic networks; see [1].

Preservation of the block structures of the data matrices also increases the accuracy of the moment-matching property of the SPRIM models. While the theory of projection onto Krylov subspace guarantees (35), the SPRIM reduced-order models satisfy the following stronger moment-matching property:

$$
H_n(s) = H(s) + \mathscr{O}\left((s-s_0)^{2\hat{q}(\hat{n},m)}\right), \quad \text{where } \hat{q}(\hat{n},m) \geq \left\lfloor \frac{\hat{n}}{m} \right\rfloor. \tag{42}
$$

The integer $\hat{q}(\hat{n},m)$ is the same as in (35) and (36), and in the generic case, $\hat{q}(\hat{n},m) = \lfloor \hat{n}/m \rfloor$ in (42). The property means that at the expansion point s_0, the transfer function of the SPRIM reduced-order model matches twice as many moments as the theory of general Krylov subspace-based projection methods predicts. The reason for this increased accuracy is a certain J-symmetry, which follows from the block structures of the data matrices of the SPRIM models and results in a doubling of the number of matched moments; for details and proofs, we refer the reader to [19].

5.4 Pros and Cons of PRIMA and SPRIM

The underlying Krylov subspace for both PRIMA and SPRIM is $\mathscr{K}_{\hat{n}}(M,R)$, and generating a basis matrix $\hat{V}_{\hat{n}}$ for this subspace is the dominant computational work for both algorithms. PRIMA uses $\hat{V}_{\hat{n}}$ directly to generate a reduced-order model of state-space dimension $\hat{n}$ that matches $\hat{q}(\hat{n},m) \geq \lfloor \hat{n}/m \rfloor$ moments and is passive.

SPRIM first processes the blocks of $\hat{V}_{\hat{n}}$ to obtain the matrix (39), V_n. This matrix is then used to generate a reduced-order model of state-space dimension n that matches $2\hat{q}(\hat{n},m) \geq \lfloor \hat{n}/m \rfloor$ moments, is passive, and satisfies reciprocity. Here, n is the number of columns of V_n. Since each of the three diagonal blocks $\tilde{V}^{(l)}$, $l = 1,2,3$, of V_n has at most $\hat{n}$ columns and $\tilde{V}^{(3)}$ has at most m_v columns, n is bounded as follows:

$$n \leq 2\hat{n} + \min\{\hat{n}, m_v\}.$$

Therefore, for the same $\hat{n}$, PRIMA generates smaller reduced-order models than SPRIM. On the other hand, the SPRIM models satisfy reciprocity, which simplifies their synthesis as electrical networks, and in terms of moment matching, the SPRIM models are twice as accurate as the PRIMA models.

In practice, both PRIMA and SPRIM are run as iterative algorithms with iteration counter $\hat{n}$. The iterative process is stopped once $\hat{n}$ is large enough so that the reduced-order transfer function has converged to the original transfer function throughout the given frequency range of interest. Due to its higher accuracy, SPRIM will typically converge faster, i.e., for a smaller $\hat{n}$, than PRIMA. The following example, which is taken from [21], illustrates the difference in convergence behavior between SPRIM and PRIMA. The example (referred to as "package example") is an RCL network with state-space dimension $N = 1841$. The network has $m_i = 8$ current sources and $m_v = 8$ voltage sources, and thus $m = m_i + m_v = 16$. Its transfer function is 16×16-matrix valued and has 256 components. The expansion point $s_0 = 2\pi \times 10^{10}$ was used. For this example, $\hat{n} = 128$ was needed for SPRIM to achieve convergence. The corresponding state-space dimension of the SPRIM reduced-order model is $n = 2\hat{n} + m_v = 264$. Figure 3 depicts the absolute values of the (8,1)-component and the (9,9)-component of the transfer functions. Note that for $\hat{n} = 128$ PRIMA has not converged yet.

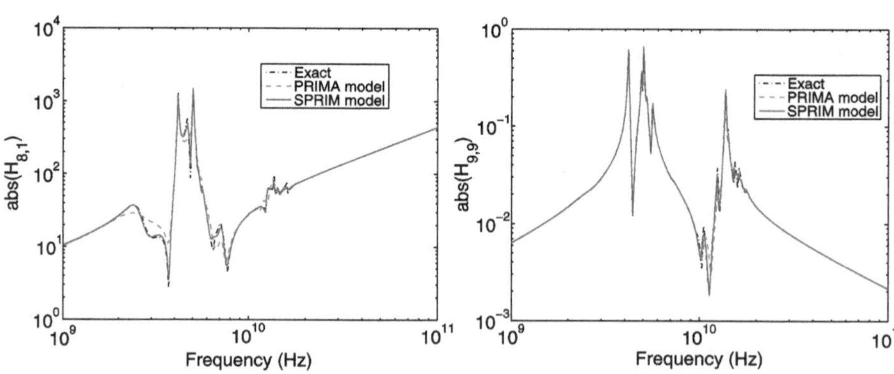

(a) (8,1)-component of transfer functions (b) (9,9)-component of transfer functions

Fig. 3 Absolute values of transfer functions for package example

6 Thick-Restart Krylov Subspace Techniques

While Krylov subspace-projection algorithms, such as PRIMA and SPRIM, generate reduced-order models that are passive and satisfy a moment-matching property, their use becomes prohibitive for RCL networks with 'very' large state-space dimension N. The computational costs of these methods is dominated by the generation of suitable basis vectors for Krylov subspaces $\mathscr{K}_n(M,R)$ of dimension n. For general RCL networks, the work for this task is proportional to n^2N. As a result, for very large N, it is typically not feasible to run the Krylov subspace method for a value of n that is large enough to produce a reduced-order model of sufficient accuracy.

A standard approach in numerical linear algebra to reduce the computational cost from $\mathscr{O}(n^2N)$ to $\mathscr{O}(n_0 lN)$ is to restart the Krylov subspace method after each cycle of n_0 steps and run l of these restarted cycles. The most common restarted Krylov subspace techniques are designed for matrix computations that are described by M and for which the starting block R can be chosen freely. In model order reduction, however, R is determined by the given data matrices of the original system, and thus R cannot be chosen freely.

Recently, there has been a lot of interest in so-called *thick-restart* Krylov subspace techniques (see, e.g., [10] and the references therein), which, in contrast to standard restart methods, can be adapted to the problem of model order reduction. The basic idea is that after each cycle of n_0 steps, 'relevant' information is extracted from the batch of basis vectors generated in this cycle and the extracted vectors are then used as starting vectors in the next cycle.

For model order reduction, 'relevant' information means converged eigenvectors corresponding to poles of the original transfer function (14), H, that are close to the frequency range of interest; cf. Fig. 2. Recall from (14) that any pole $\mu \in \mathbb{C}$ of H is a generalized eigenvalue of the matrix pencil $sE - A$; i.e., there exists a generalized eigenvector $v \in \mathbb{C}^N$ such that

$$Av = \mu Ev, \quad v \neq 0. \tag{43}$$

The Krylov subspace method is applied to the matrix $M = -(s_0 E - A)^{-1}E$. It is straightforward to verify that the vector v satisfies the relation (43) if, and only if, v is an eigenvector of M, i.e.,

$$Mv = \lambda v, \quad v \neq 0, \tag{44}$$

where the eigenvalues μ and λ are connected as follows:

$$\mu = s_0 + \frac{1}{\lambda}. \tag{45}$$

It is well known that Krylov subspace methods (applied to M) will generate eigenvectors for the dominant eigenvalues λ of M within a very small number of iterations. Here, 'dominant' means 'largest in absolute value'. In view of (45), the dominant eigenvalues λ of M correspond to the poles μ of H closest to the expansion point s_0. This means that by placing s_0 close to the frequency range of interest, we can expect to obtain generalized eigenvectors associated with a few converged poles of H close to the frequency range of interest. These generalized eigenvectors are then used for the thick restart in the next cycle of n_0 steps of the Krylov subspace method. The same process is repeated in such a way that in each cycle, one obtains additional converged poles of H close to the frequency range of interest. However, since poles closest to the expansion point converge fastest, this thick-restart process has to allow for changing expansion points, so that in each cycle new additional poles of H can be found. We denote by $s_0^{(j)}$ the expansion point employed in the j-th cycle. Note that, except for the first point $s_0^{(1)}$, the expansion points need to be complex in order to stay close to the frequency range of interest; cf. Fig. 4. A typical strategy for choosing the $s_0^{(j)}$'s is to move up parallel to the imaginary axis, as indicated in Fig. 4 for the case $l = 3$.

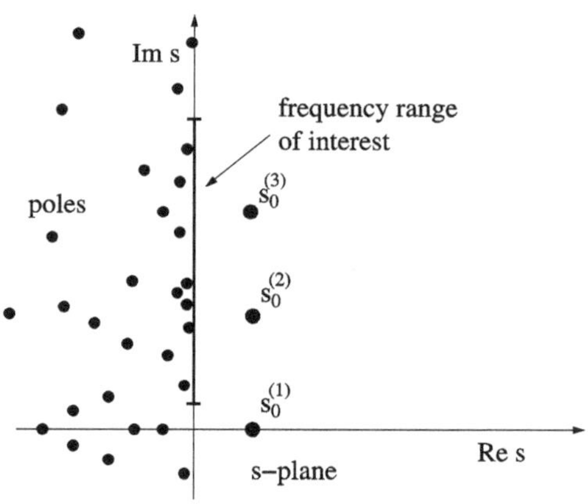

Fig. 4 Placement of multiple expansion points $s_0^{(1)}, s_0^{(2)}, \ldots, s_0^{(l)}$

With changing values of $s_0^{(j)}$, the matrix M also changes within each cycle. We denote by

$$M^{(j)} = -\left(s_0^{(j)} E - A\right)^{-1} E$$

the matrix that the Krylov subspace method is applied to in the l-th cycle. It readily follows from the equivalence of (44) and (43) that any eigenvector v of $M^{(j)}$ is also an eigenvector of $M^{(j+1)}$, only the value of the corresponding eigenvalues are transformed according to the relation

$$s_0^{(j)} + \frac{1}{\lambda^{(j)}} = s_0^{(j+1)} + \frac{1}{\lambda^{(j+1)}}.$$

Due to this invariance of the eigenvectors of $M^{(j)}$, the information used in the thick restart remains relevant from cycle to cycle even though the matrix $M^{(j)}$ changes to $M^{(j+1)}$.

In the following algorithm, we outline the basic steps for generating a projection matrix V_n via the thick-restart Krylov subspace approach.

Algorithm 5 (Generation of projection matrix V_n via thick-restart Krylov subspace approach)

- ***Input:*** *Matrices of the form*

$$A = \begin{bmatrix} A_{11} & -\mathscr{A}_l & -\mathscr{A}_v \\ \mathscr{A}_l^T & 0 & 0 \\ \mathscr{A}_v^T & 0 & 0 \end{bmatrix}, \qquad E = \begin{bmatrix} E_{11} & 0 & 0 \\ 0 & L & 0 \\ 0 & 0 & 0 \end{bmatrix}, \qquad B = \begin{bmatrix} \mathscr{A}_i & 0 \\ 0 & 0 \\ 0 & -I \end{bmatrix},$$

  where $A_{11} \preceq 0$, $E_{11} \succeq 0$, *and* $L \succ 0$.
  An initial expansion point $s_0^{(1)} \in \mathbb{R}$ *such that the matrix* $s_0^{(1)} E - A$ *is nonsingular; this last condition is satisfied for any* $s_0^{(1)} > 0$.
  Number n_0 $(> m)$ *of iterations in each cycle of the Krylov subspace method.*
  Maximum number $l_{\max}$ *of thick restarts.*
- *For* $l = 1, 2, \ldots, l_{\max}$ *do:*
  - *Formally set*

$$M^{(l)} = -\left(s_0^{(l)} E - A\right)^{-1} E, \qquad R^{(l)} = \begin{cases} (s_0^{(l)} E - A)^{-1} B & \text{if } l = 1, \\ Y^{(l-1)} & \text{if } l > 1. \end{cases}$$

  - *Run* n_0 *steps of your favorite Krylov subspace method (applied to* $M^{(l)}$ *and* $R^{(l)}$*) to obtain an* $n \times n_0$ *matrix* $V^{(l)}$ *the columns of which span the Krylov subspace* $\mathscr{K}_{n_0}(M^{(l)}, R^{(l)})$.
  - *If* l *is 'large' enough: stop.*
  - *Extract 'relevant' eigenvector information* $Y^{(l)}$ *from* $V^{(l)}$ *to be used in the next thick restart.*
  - *Select suitable next expansion point* $s_0^{(l)} \in \mathbb{C}$. *It needs to be such that the matrix* $s_0^{(l)} E - A$ *is nonsingular; this last condition is satisfied if* $\operatorname{Re} s_0^{(l)} > 0$.
- ***Output:*** *Projection matrix*

$$V_n = \left[V^{(1)}\ V^{(2)}\ \cdots\ V^{(l)}\right], \quad \text{where } n := n_0 l. \tag{46}$$

The matrix V_n produced by Algorithm 5 is then used to obtain a reduced-order model of state-space dimension n via projection by defining the reduced-order data matrices as follows:

$$A_n := V_n^H A V_n, \qquad E_n := V_n^H E V_n, \quad \text{and} \quad B_n := V_n^H B. \tag{47}$$

Note that due to complex expansion points, the matrix V_n is complex in general, and thus A_n, E_n, and B_n are complex in general as well. In Sect. 7 below, we briefly discuss how to obtain real projection matrices even for complex expansion points.

Due to the use of the multiple expansion points $s_0^{(1)}, s_0^{(2)}, \ldots, s_0^{(l)}$, the reduced-order model given by the data matrices (47) satisfies a multi-point moment-matching property of the form

$$H_n(s) = H(s) + \mathcal{O}\big(\big(s - s_0^{(j)}\big)^{q_j}\big), \quad j = 1, 2, \ldots, l. \tag{48}$$

We remark that reduced-order models characterized by (48) can also be obtained by complex frequency hopping (CFH) [7]. However, CFH was proposed as a modification of AWE to allow changing expansion points, and CFH suffers from the same numerical problems as AWE due to the use of explicit moment computations.

Finally, we present a numerical example that illustrates the benefits of thick-restart Krylov subspace techniques for reduced-order modeling. With a single real expansion point, a reduced-order model of state-space dimension $n = 80$ is needed to achieve satisfactory convergence throughout the frequency range of interest; see Fig. 5a. With thick restarts, three expansion points (one real, two complex) and $l = 3$ cycles of $n_0 = 14$ Krylov stops are sufficient to obtain a reduced-order model of state-space dimension $n = n_0 l = 42$ of comparable accuracy; see Fig. 5b.

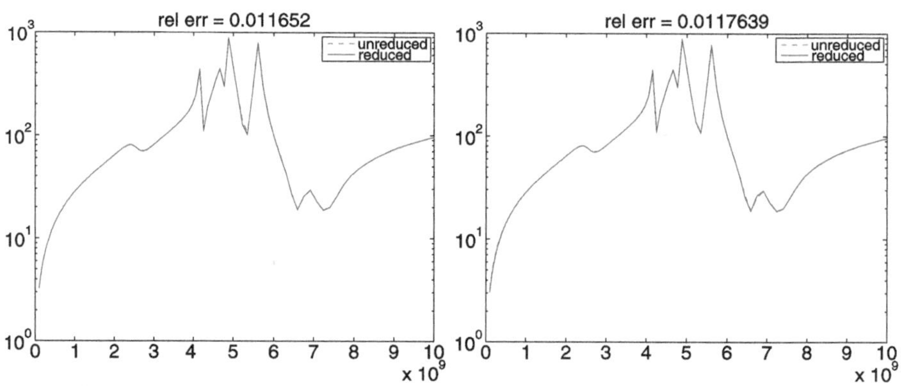

(a) Single expansion point and $n = 80$ (b) Three expansion points and $n = 42$

Fig. 5 Single expansion point vs. multiple expansion points and thick restarts

7 Complex Expansion Points

There are two main reasons why Krylov subspace-based model order reduction usually employs real expansion points s_0. First, the matrices (28), M and R, to which the Krylov subspace method is applied are real, thus avoiding the use of complex arithmetic, which is 4 times as costly as real arithmetic. Second, when passive reduced-order models are constructed via the projection approach described in Sect. 3.4, the projection matrix V_n needs to be real in order to obtain real reduced-order data matrices (33). On the other hand, the use of complex expansion points s_0 typically results in a significantly smaller state-space dimension n of the reduced-order models, since s_0 can be placed closer to the frequency range of interest than any real s_0. Finally, as we discussed in Sect. 6, thick-restart Krylov subspace techniques only make sense when complex expansion points can be used.

We now consider the case that $s_0 \in \mathbb{C} \setminus \mathbb{R}$ is a non-real expansion point and that $V_n \in \mathbb{C}^{N \times n}$ is a basis matrix for the complex n-dimensional Krylov subspace $\mathcal{K}_n(M,R)$, where M and R are the complex matrices given by (28). Suppose we are trying to employ a projection approach similar to the one in Sect. 3.4 that produces real reduced-order data matrices. One possibility is to replace the complex matrix V_n by the real matrix

$$\begin{bmatrix} \operatorname{Re} V_n & \operatorname{Im} V_n \end{bmatrix}; \tag{49}$$

see [26, 27]. One obvious disadvantage of this approach is that the dimension of the resulting reduced-order model is doubled to $2n$. Furthermore, in general, the matrix (49) is not guaranteed to have full column rank, and so before using (49) as a projection matrix, one would need to check for and possibly delete any linearly dependent columns of (49) by means of some variant of a Gram-Schmidt orthogonalization process. On the other hand, the transfer function of the resulting reduced-order model will satisfy a two-point moment-matching property of the form

$$H_n = H(s) + \mathcal{O}\big((s-s_0)^{\hat{q}(n,m)}\big) \quad \text{and} \quad H_n = H(s) + \mathcal{O}\big((s-\overline{s_0})^{\hat{q}(n,m)}\big), \tag{50}$$

where $\hat{q}(n,m) \geq \lfloor n/m \rfloor$.

It turns out that the process of generating a real projection matrix by first computing a complex basis matrix V_n for $\mathcal{K}_n(M,R)$ and then orthogonalizing the columns of the matrix (49) is computationally inefficient. First note that the resulting real projection matrix is a basis matrix for the *n-th paired complex conjugate Krylov subspace*

$$\mathcal{K}_n^{(\mathrm{p})}(M,R) := \operatorname{span}\big\{\, v, \overline{v} \;\big|\; v \in \mathcal{K}_n(M,R) \,\big\}. \tag{51}$$

In [24], we study the problem of constructing real basis matrices for paired complex conjugate Krylov subspaces $\mathcal{K}_n^{(\mathrm{p})}(M,R)$ and propose an algorithm that is computationally cheaper than the obvious approach outlined above. In particular, employing the algorithm from [24] allows the efficient construction of passive reduced-order models via projection onto Krylov subspaces with non-real expansion points.

Finally, we present a numerical example that illustrates the benefits of using even a single complex expansion point. With a single real expansion point, a reduced-order model of state-space dimension $n = 138$ is needed to achieve satisfactory convergence throughout the frequency range of interest; see Fig. 6a. With a single complex expansion point, a reduced-order model of state-space dimension $n = 69$ is sufficient to obtain comparable accuracy; see Fig. 6b. Thus the dimension of the reduced-order model has been halved by employing a complex expansion point.

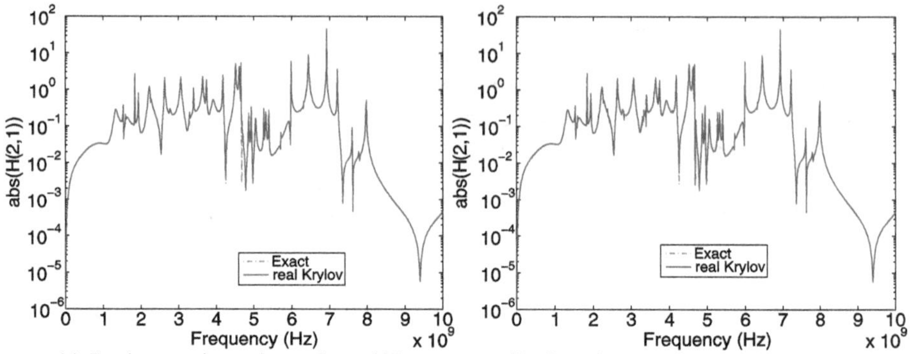

(a) Real expansion point and $n = 138$

(b) Complex expansion point and $n = 69$

Fig. 6 Single real vs. single complex expansion point

8 Concluding Remarks

Model order reduction is a classical and well-studied subject of control theory. However, in control theory, the systems to be reduced are of relatively small or at most moderate size, and most of the methods developed in this area are not suitable or efficient enough to be applied to large-scale systems. In the early 1990s, the need to efficiently reduce the ever-increasing sizes of the RCL networks used to model the interconnect of VLSI circuits led to a renewed interest in model order reduction especially in the development of methods that are applicable to large-scale systems. Model order reduction techniques based on Krylov subspaces have emerged as the methods of choice for generating macromodels of large-scale multi-port RCL networks that arise in VLSI circuit simulation. Furthermore, these algorithms have also found applications in other areas, such as structural analysis and computational acoustics.

Despite all the progress in Krylov subspace-based model order reduction of large-scale RCL networks in recent years, there are still many open problems. State-of-the-art structure-preserving methods, such as SPRIM, first generate a basis matrix of the underlying Krylov subspace and then employ explicit projection using some suitable partitioning of the basis matrix to obtain a structure-preserving reduced-order model. In particular, there are two major problems with the use of such explicit projections. First, it requires the storage of the basis matrix, which becomes prohibitive in the case of 'truly' large-scale RCL networks. Second, the approximation properties of the resulting structure-preserving reduced-order models are not optimal, and they show that the available degrees of freedom are not fully used in general. It would be highly desirable to have structure-preserving model order reduction methods that do not involve explicit projection and would thus be applicable in the truly large-scale case. Other unresolved issues include the automatic and adaptive choice of the expansion points and robust and reliable stopping criteria and error bounds.

Acknowledgements This work was supported in part by the National Science Foundation through Grant DMS-0613032.

References

1. Anderson, B.D.O., Vongpanitlerd, S.: Network Analysis and Synthesis. Prentice-Hall, Englewood Cliffs, New Jersey (1973)
2. Antoulas, A.C.: Approximation of Large-Scale Dynamical Systems. SIAM Publications, Philadelphia, Pennsylvania (2005)
3. Antoulas, A.C., Sorensen, D.C.: Approximation of large-scale dynamical systems: An overview. Int. J. Appl. Math. Comput. Sci. **11**, 1093–1121 (2001)
4. Bai, Z., Feldmann, P., Freund, R.W.: How to make theoretically passive reduced-order models passive in practice. In: Proc. IEEE 1998 Custom Integrated Circuits Conference, pp. 207–210. IEEE, Piscataway, New Jersey (1998)
5. Baker, Jr., G.A., Graves-Morris, P.: Padé Approximants, second edn. Cambridge University Press, New York, New York (1996)
6. Celik, M., Pileggi, L., Odabasioglu, A.: IC Interconnect Analysis. Kluwer Academic Publishers, Boston/Dordrecht/London (2002)
7. Chiprout, E., Nakhla, M.S.: Analysis of interconnect networks using complex frequency hopping (CFH). IEEE Trans. Computer-Aided Design **14**(2), 186–200 (1995)
8. de Villemagne, C., Skelton, R.E.: Model reductions using a projection formulation. Internat. J. Control **46**(6), 2141–2169 (1987)
9. Deo, N.: Graph Theory with Applications to Engineering and Computer Science. Prentice-Hall, Englewood Cliffs, New Jersey (1974)
10. Eiermann, M., Ernst, O.G.: A restarted Krylov subspace method for the evaluation of matrix functions. SIAM J. Numer. Anal. **44**, 2481–2504 (2006)
11. Elmore, W.C.: The transient response of damped linear networks with particular regard to wideband amplifiers. J. Appl. Phys. **19**(1), 55–63 (1948)
12. Feldmann, P., Freund, R.W.: Efficient linear circuit analysis by Padé approximation via the Lanczos process. In: Proceedings of EURO-DAC '94 with EURO-VHDL '94, pp. 170–175. IEEE Computer Society Press, Los Alamitos, California (1994)
13. Feldmann, P., Freund, R.W.: Efficient linear circuit analysis by Padé approximation via the Lanczos process. IEEE Trans. Computer-Aided Design **14**, 639–649 (1995)
14. Feldmann, P., Freund, R.W.: Reduced-order modeling of large linear subcircuits via a block Lanczos algorithm. In: Proc. 32nd ACM/IEEE Design Automation Conference, pp. 474–479. ACM, New York, New York (1995)
15. Freund, R.W.: Computation of matrix Padé approximations of transfer functions via a Lanczos-type process. In: C. Chui, L. Schumaker (eds.) Approximation Theory VIII, Vol. 1: Approximation and Interpolation, pp. 215–222. World Scientific Publishing Co., Inc., Singapore (1995)
16. Freund, R.W.: Krylov-subspace methods for reduced-order modeling in circuit simulation. J. Comput. Appl. Math. **123**(1–2), 395–421 (2000)
17. Freund, R.W.: Model reduction methods based on Krylov subspaces. Acta Numerica **12**, 267–319 (2003)
18. Freund, R.W.: SPRIM: structure-preserving reduced-order interconnect macromodeling. In: Tech. Dig. 2004 IEEE/ACM International Conference on Computer-Aided Design, pp. 80–87. IEEE Computer Society Press, Los Alamitos, California (2004)
19. Freund, R.W.: On Padé-type model order reduction of J-Hermitian linear dynamical systems. Linear Algebra Appl. **429**, 2451–2464 (2008)
20. Freund, R.W.: Structure-preserving model order reduction of RCL circuit equations. In: W. Schilders, H. van der Vorst, J. Rommes (eds.) Model Order Reduction: Theory, Research

Aspects and Applications, Mathematics in Industry, vol. 13, pp. 49–73. Springer-Verlag, Berlin/Heidelberg, Germany (2008)
21. Freund, R.W.: The SPRIM algorithm for structure-preserving order reduction of general RCL circuits. In: P. Benner, M. Hinze, J. ter Maten (eds.) Model Reduction for Circuit Simulation, Lecture Notes in Electrical Engineering. Springer-Verlag, Berlin/Heidelberg, Germany (2011). To appear
22. Freund, R.W., Feldmann, P.: The SyMPVL algorithm and its applications to interconnect simulation. In: Proc. 1997 International Conference on Simulation of Semiconductor Processes and Devices, pp. 113–116. IEEE, Piscataway, New Jersey (1997)
23. Freund, R.W., Feldmann, P.: Reduced-order modeling of large linear passive multi-terminal circuits using matrix-Padé approximation. In: Proc. Design, Automation and Test in Europe Conference 1998, pp. 530–537. IEEE Computer Society Press, Los Alamitos, California (1998)
24. Freund, R.W., Rensi, E.B.: Computing orthonormal bases of paired complex conjugate Krylov subspaces. Technical report, Department of Mathematics, University of California, Davis, California (2011). In preparation
25. Grimme, E.J.: Krylov projection methods for model reduction. Ph.D. thesis, Department of Electrical Engineering, University of Illinois at Urbana-Champaign, Urbana-Champaign, Illinois (1997)
26. Kamon, M., Marques, N.A., Silveira, L.M., White, J.: Automatic generation of accurate circuit models of 3-D interconnect. IEEE Trans. Compon. Packag. Manuf. Technol.—Part B **21**(3), 225–240 (1998)
27. Lassaux, G., Willcox, K.E.: Model reduction for active control design using multiple-point Arnoldi methods. In: Proceedings of the 41st Aerospace Sciences Meeting & Exhibit. Reno, Nevada (2003). Paper AIAA-2003-0616
28. Odabasioglu, A.: Provably passive RLC circuit reduction. M.S. thesis, Department of Electrical and Computer Engineering, Carnegie Mellon University (1996)
29. Odabasioglu, A., Celik, M., Pileggi, L.T.: PRIMA: passive reduced-order interconnect macromodeling algorithm. In: Tech. Dig. 1997 IEEE/ACM International Conference on Computer-Aided Design, pp. 58–65. IEEE Computer Society Press, Los Alamitos, California (1997)
30. Odabasioglu, A., Celik, M., Pileggi, L.T.: PRIMA: passive reduced-order interconnect macromodeling algorithm. IEEE Trans. Computer-Aided Design **17**(8), 645–654 (1998)
31. Pillage, L.T., Rohrer, R.A.: Asymptotic waveform evaluation for timing analysis. IEEE Trans. Computer-Aided Design **9**, 352–366 (1990)
32. Pillage, L.T., Rohrer, R.A., Visweswariah, C.: Electronic circuit and system simulation methods. McGraw-Hill, Inc., New York, New York (1995)
33. Ruehli, A.E. (ed.): Circuit Analysis, Simulation, and Design Part 1: General Aspects of Circuit Analysis and Design. North-Holland, Amsterdam, The Netherlands (1986)
34. Sheehan, B.N.: ENOR: model order reduction of RLC circuits using nodal equations for efficient factorization. In: Proc. 36th ACM/IEEE Design Automation Conference, pp. 17–21. ACM, New York, New York (1999)
35. Vlach, J., Singhal, K.: Computer Methods for Circuit Analysis and Design, second edn. Van Nostrand Reinhold, New York, New York (1994)
36. Willems, J.C.: Dissipative dynamical systems, part ii: Linear systems with quadratic supply rates. Arch. Rational Mech. Anal. **45**, 352–393 (1972)

Injection Locking Analysis and Simulation of Weakly Coupled Oscillator Networks

Prateek Bhansali and Jaijeet Roychowdhury

Abstract Coupled oscillator networks (CONs) occur in various domains such as biology and electronics. The nonlinear phenomenon of "injection locking" leads to frequency synchronization among the oscillators in a CON. In this chapter, we first obtain simplified tools to analyze the phenomenon of injection locking and present the Generalized Adler's equation. The Generalized Adler's equation is applicable for analyzing the injection locking in any type of oscillator. We demonstrate the use of Generalized Adler's equation for injection locking analysis on a ring oscillator and a Hodgkin-Huxley neuronal oscillator. Next, we describe efficient system-level simulation techniques (synchronized steady state and transient) for coupled oscillator networks using the Perturbation Projection Vector (PPV) phase macromodel of the oscillator. The coupled oscillator simulation techniques are generic in nature and can be applied for the simulation of any coupled oscillator network, like the ones found in biological or electronic systems. We apply our system-level techniques on a $-G_m$ LC (harmonic) and a Hodgkin-Huxley (non-harmonic) CON.

1 Introduction

Coupled oscillator networks are found throughout the physical and biological worlds as sub-systems, often interacting in a complex fashion with other parts of the larger system. The interaction between individual oscillators in a CON can result in several interesting behaviors, notably the process of synchronization through a phenomenon called injection locking [1, 2]. For example, each sinoatrial node found in the natural pacemaker of the heart acts as an oscillator generating periodic electrical signals. The reliable periodic pumping of the heart is driven by rhythmic electri-

Prateek Bhansali
University of California, Berkeley, USA, e-mail: bhansali@berkeley.edu

Jaijeet Roychowdhury
University of California, Berkeley, USA, e-mail: jr@berkeley.edu

P. Li et al. (eds.), *Simulation and Verification of Electronic and Biological Systems*,
DOI 10.1007/978-94-007-0149-6_4, © Springer Science+Business Media B.V. 2011

cal activity of the synchronized network of these pacemaker cells. The synchronous flashing of fireflies is another example of coupled oscillators in action in nature. In electrical engineering, coupled oscillator networks are used for practical applications such as low power global clock distribution, and synthesis of beam scanning arrays.

As noted above, owing to its practical and theoretical importance, the analysis of CON dynamics has received much attention in research [1, 3–5]. Each individual oscillator in a CON is mathematically modeled by a set of nonlinear, coupled differential equations. Solving such coupled differential equations analytically, even for a single oscillator, is almost always a difficult task. Thus, automated analysis and simulation of coupled oscillators is necessarily required. The direct time-course integration method is not suitable for oscillator related simulations as it inherently accumulates phase errors without limit during simulation [6]. Thus, to obtain acceptable accuracy using direct time-course methods, hundreds of timesteps per oscillation cycle will be required, leading to higher computational costs for a full system simulation. It is in this context we propose efficient system-level simulation techniques for coupled oscillator networks using the PPV phase macromodel of the oscillator [6]. These coupled oscillator simulation techniques are generic in nature and can be used for the simulation of any biological or electronic CONs.

We first obtain simplified tools to analyze the phenomenon of injection locking which causes the rich and complex dynamical behavior of CONs. Injection locking occurs when a small periodic injection signal of frequency within the "locking range" entrains the oscillator at its frequency. Upon entrainment the oscillator is said to be locked to the injection signal and this phenomenon is known as the injection locking. The prevalent technique to analyze injection locking is to use classical Adler's equation [2]. However, injection locking analysis based on classical Adler's equation is limited to LC oscillators (harmonic oscillators) only, because it explicitly depends on the quality factor, a concept inherent to LC oscillators. Moreover, classical Adler's equation assumes a sinusoidal injection signal, which limits its general applicability. We present the Generalized Adler's equation applicable for injection locking analysis on oscillators independent of its type (harmonic or non-harmonic). This equation is obtained by averaging the PPV phase macromodel. The procedure is considerably simple and handy in determining the locking range for any arbitrarily shaped weak periodic injection signal.

However, it is not possible to obtain simplified tools for complex coupled oscillator systems while avoiding approximations (such as averaging). We describe an efficient and accurate transient simulation technique for CONs based on the PPV phase macromodel of the oscillator. The PPV phase macromodel abstracts the behavior of an oscillator accurately in a single variable irrespective of the original oscillator state size, and hence can be used for system level design and efficient simulation. We also develop an efficient finite-difference time-domain (FDTD) technique for computing the synchronized state phase distribution and synchronized frequency of the oscillators directly from the phase domain description of a CON instead of original system equations.

The remainder of this chapter is organized as follows. In Sect. 2, we review the mathematical background of a single oscillator and the PPV phase macromodel of an oscillator. Next, in Sect. 3, we provide an overview of injection locking and obtain the Generalized Adler's equation for its analysis. In Sect. 4, we provide a recipe for transient simulation of CONs using the PPV phase macromodel. Finally, we describe the theoretical formulation and the numerical method in detail for computing the synchronized steady state of a CON starting with its phase domain description.

2 Oscillators

A nonlinear oscillator perturbed by an external input $\mathbf{b}(t)$ can be described by the following set of n Differential-Algebraic Equations (DAEs)

$$\frac{d\mathbf{q}(\mathbf{x}(t))}{dt} = \mathbf{r}(\mathbf{x}(t)) + \mathbf{b}_{\text{ext}}(t). \tag{1}$$

In the above equation $t \in \mathbb{R}$ represents the time, and $\mathbf{x}(\cdot) : \mathbb{R} \mapsto \mathbb{R}^n$ represents the oscillator state variables. Next, $\mathbf{r}(\cdot) : \mathbb{R}^n \mapsto \mathbb{R}^n$ and $\mathbf{q}(\cdot) : \mathbb{R}^n \mapsto \mathbb{R}^n$ represents certain nonlinear function depending on the oscillator type and topology. In the context of circuits, $\mathbf{r}(\cdot)$ and $\mathbf{q}(\cdot)$ represents the contributions of memoryless nonlinearities and charge/flux elements, respectively. Finally, $\mathbf{b}_{\text{ext}}(\cdot) : \mathbb{R} \mapsto \mathbb{R}^n$ represents external perturbations to the oscillator circuit. In the absence of external perturbation ($\mathbf{b}_{\text{ext}}(t) = 0$), the oscillator exhibits a T-periodic solution for DAEs

$$\frac{d\mathbf{q}(\mathbf{x}(t))}{dt} = \mathbf{r}(\mathbf{x}(t)) \tag{2}$$

in the steady state such that $\mathbf{x}_{ss}(t) = \mathbf{x}_{ss}(t+T)$. For example, Fig. 1 shows a three stage ring oscillator perturbed by an external injection signal $b(t)$. The DAEs describing such a ring oscillator are

$$\begin{aligned}
C\frac{dv_1}{dt} &= -\frac{v_1(t)}{R} + \frac{f(v_3)}{R} \\
C\frac{dv_2}{dt} &= -\frac{v_2(t)}{R} + \frac{f(v_1)}{R} \\
C\frac{dv_3}{dt} &= -\frac{v_3(t)}{R} + \frac{f(v_2)}{R} - b(t),
\end{aligned} \tag{3}$$

where $f(v_i) = \tanh(G_m v_i(t))$ models the inverter characteristics, and $v_i(t)$ represents node voltages. In the absence of external injection signal ($b(t) = 0$) the oscillator node voltages settles down to a Periodic Steady State (PSS). For $R = 1$ kΩ, $C = 2$ pF, and $G_m = -50$ the PSS of ring oscillator is shown in the Fig. 2.

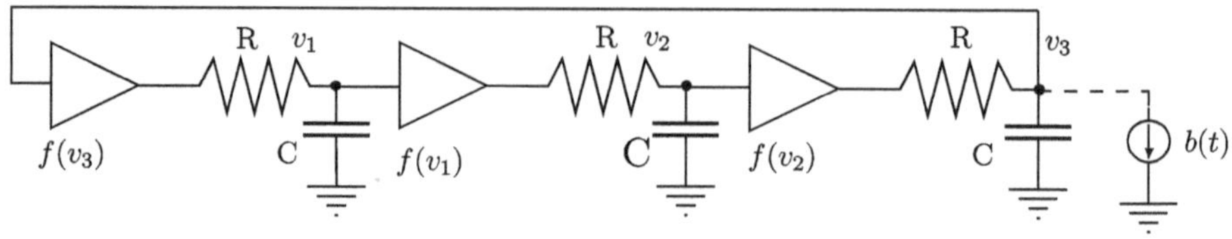

Fig. 1 A three stage ring oscillator perturbed with an injection signal $b(t)$

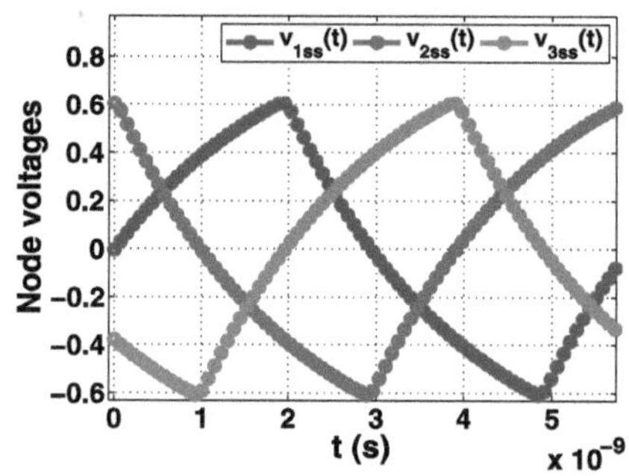

Fig. 2 PSS ($\mathbf{x}_{ss}(t)$) of the three stage ring oscillator, $T = 5.78 \times 10^{-9}$ s ($f = 1.73 \times 10^{8}$ Hz)

2.1 PPV/PRC Phase Macromodel of Single Oscillator

If the external perturbation, $\mathbf{b}_{ext}(t)$, is small then it can be shown that the solution to (1) can be computed from the periodic steady state solution as

$$\mathbf{x}(t) = \mathbf{x}_{ss}(t + \alpha(t)), \tag{4}$$

where $\alpha(t) : \mathbb{R} \mapsto \mathbb{R}$ is known as the phase shift due to external perturbations and $\mathbf{x}_{ss}(t)$ is the steady state T-periodic solution in the absence of external perturbations [6]. Furthermore, $\alpha(t)$ is given by the phase shift equation

$$\frac{d\alpha(t)}{dt} = \mathbf{v}_1^T(t + \alpha(t)) \cdot \mathbf{b}_{ext}(t), \tag{5}$$

where $\mathbf{v}_1^T(\cdot) : \mathbb{R} \mapsto \mathbb{R}^n$ is known as the PPV (Perturbation Projection Vector) of the oscillator. The PPV represents phase sensitivities of the oscillator state variables to the external perturbations. It is a T-periodic vector and can be computed numerically using either the Floquet theory based approach [6, 7] or Malkin's theorem [8]. The PPV is well known to the biological community as the Phase Response Curve (PRC) and is used extensively for circadian rhythms and neuronal oscillators analysis [1, 8].

2.2 Malkin's Theorem

Here, we informally state the Malkin's theorem. If the unperturbed oscillator in (2) is asymptotically orbitally stable [9] then its phase variation under the effect of an external perturbation as in (1) is described by (5) and the T-periodic PPV (or PRC) $\mathbf{v}_1(t)$ is given by the solution of the Linear Periodically Time-Varying (LPTV) adjoint equation

$$\frac{d\mathbf{v}_1(t)}{dt} = -G(t)^T \mathbf{v}_1(t) \tag{6}$$

with the normalization condition $\mathbf{v}_1^T(0) \cdot \mathbf{r}(\mathbf{x}_{ss}(0)) = f$, where $G(t)_{n\times n} = \frac{\partial \mathbf{r}(\mathbf{x}_{ss}(t))}{\partial \mathbf{x}_{ss}(t)}$ and f is oscillator's free running frequency [8]. Note that the availability of the oscillator's periodic steady state solution is critical for the evaluation of its PPV algorithmically as in (6).

The PPV waveforms, for the ring oscillator shown in Fig. 1, computed using the Malkin's theorem are shown in Fig. 3a. Note that the PPV waveforms are periodic with the same time period as the oscillator's steady state. We now perturb the ring

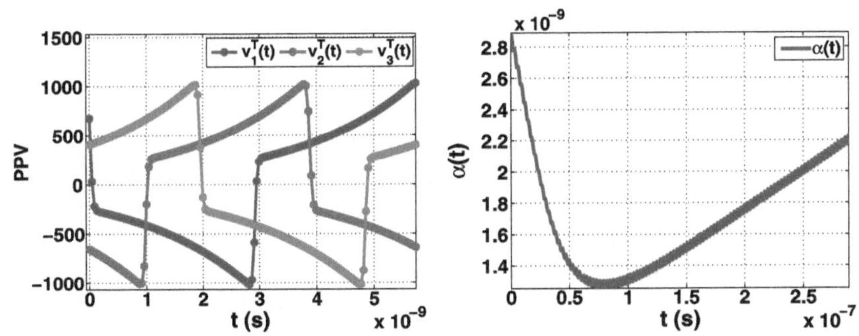

(a) PPV waveforms (b) $\alpha(t)$ for $b(t) = 10^{-04}\sin(2\pi 1.74\times 10^8 t)$.

Fig. 3 PPV and $\alpha(t)$ of the ring oscillator

oscillator with the injection signal, $b(t) = 10^{-4}\sin(2\pi 1.74 \times 10^8 t)$, applied at v_3 as shown in Fig. 1. The perturbation signal causes the oscillator to deviate from its steady state trajectories. We can use the PPV phase macromodel to study the transient behavior of the oscillator under perturbation. The phase deviation caused by this injection signal can be computed using (5) and is shown in Fig. 3b. Using the phase deviation, $\alpha(t)$, the perturbed solution can now be computed from (4) as

$$v_i(t) = v_{iss}(t + \alpha(t)), \quad \forall i = 1,2,3. \tag{7}$$

We show the perturbed solution obtained using full simulation of (3) and compare it with the one obtained using (7) in Fig. 4.

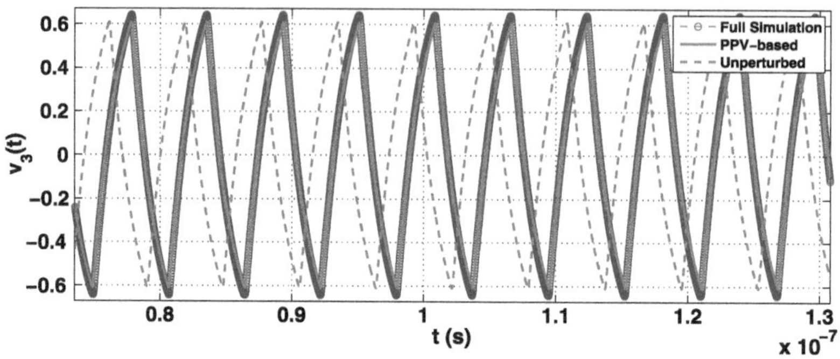

Fig. 4 Perturbed solution using the PPV phase macromodel of the three stage ring oscillator (only $v_3(t)$ is shown)

In the next section, we provide an overview of the classical Adler's equation which is extensively used for injection locking analysis. Next, we obtain Adler-like equation called as Gen-Adler using the PPV phase macromodel.

3 Injection Locking

Injection locking is a nonlinear phenomenon observed in oscillators. A small injection signal of frequency within the "locking range" entrains the oscillator at its frequency. The oscillator is thus said to be locked and the phenomenon is known as the injection locking. The synchronization among the oscillators in a coupled oscillator network is the result of injection locking process. Injection locking is fruitfully exploited in the frequency synthesis and related applications in electronics. For example, it is used in the design of high performance quadrature oscillators [10, 11] and injection locked phase-locked loops (PLLs) [12].

Injection locking has been widely studied for LC-tank based oscillators [13] using the classical Adler's equation [2]. Adler's equation provides quick insight into the locking dynamics. Furthermore, it provides rapid estimation of the locking range of oscillators for a given injection signal. The locking range is the frequency range around the oscillator's natural frequency for which injection locking occurs. But, Adler's equation is dependent on quality factor Q, which limits its use to LC oscillators. Furthermore, the Adler's equation is only applicable for sinusoidal injection signals. It is in this context, we first briefly review the classical Adler's equation [2], and derive the Generalized Adler's (Gen-Adler's) equation from the PPV nonlinear phase macromodel. We apply an averaging technique on the PPV, to average out "fast" varying behavior and retain "slow" varying behavior, obtaining Gen-Adler's. It is worthwhile to emphasize that Gen-Adler's equation can be obtained for any oscillator with arbitrary injection signal of small amplitude unlike the classical Adler's equation.

3.1 Adler's Equation

In 1946, R. Adler obtained a differential equation known as Adler's equation to model the phase difference dynamics between the LC oscillator and its injection signal. When a free running LC oscillator is perturbed by a sinusoidal injection signal, the phase difference dynamics are given by

$$\frac{d\Delta\phi(t)}{dt} = \Delta f_0 - \frac{V_i}{V}\frac{f_0}{2Q}\sin(\Delta\phi(t)), \tag{8}$$

where $\Delta\phi(t)$ is the phase difference. The amplitude of the injected signal is V_i and the output amplitude of the oscillator is V while the oscillator runs at a free running frequency of f_0. The frequency difference between the injected signal and the oscillator is Δf_0. Adler obtained injection locking behavior and the locking range for LC oscillators based on (8). When the oscillator is locked, the phase difference becomes constant and thus (8) reduces to the algebraic equation

$$\frac{\Delta f_0}{f_0} = \frac{1}{2Q}\frac{V_i}{V}\sin(\Delta\phi(t)). \tag{9}$$

Since $\sin(\cdot)$ lies between $+1$ and -1, we have

$$-\frac{1}{2Q}\frac{V_i}{V} \leq \frac{\Delta f_0}{f_0} \leq \frac{1}{2Q}\frac{V_i}{V}. \tag{10}$$

This immediately gives the locking range f_L as

$$\begin{aligned} f_L &= 2|\Delta f_0|_{\max} \\ &= \frac{f_0}{Q}\frac{V_i}{V}. \end{aligned} \tag{11}$$

3.2 Generalized Adler's Equation

In this section, we use the PPV phase macromodel reviewed in Sect. 2 to obtain the Generalized Adler's equation. We first change the variables in (5) from phase deviation $\alpha(t)$ to phase difference $\Delta\phi(t)$ to obtain a phase difference equation. We then perform averaging in the phase equation retaining the slow behavior to derive the Gen-Adler's.

Assuming that the unperturbed oscillator's frequency is f and rewriting its PPV in the form $\mathbf{v}_1^T(t) = \upsilon(ft)$, where $\upsilon(\cdot)$ is 1-periodic, we can rewrite the phase equation as

$$\frac{d\theta(t)}{dt} = f + f\upsilon(\theta(t))\cdot b(t), \tag{12}$$

where $\theta(t)$ is the phase of the oscillator and is given by $\theta(t) = f(t+\alpha(t))$. Also assuming the perturbation signal is periodic with frequency f_1, the above equation is of the following form

$$\frac{d\alpha(t)}{dt} = \upsilon(f_0(t+\alpha(t)))\cdot b(f_1 t), \tag{13}$$

where $b(\cdot)$ is 1-periodic function now. We can define phase difference between the perturbation signal and the oscillator as $\Delta\phi(t) = \theta(t) - \theta_1(t)$, for $\theta(t) = f_0(t+\alpha(t))$ and $\theta_1(t) = f_1 t$, to obtain

$$\begin{aligned} \Delta\phi(t) &= f_0(t+\alpha(t)) - f_1 t, \quad \text{or} \\ \alpha(t) &= \frac{\Delta\phi(t)}{f_0} + \frac{f_1 - f_0}{f_0} t. \end{aligned} \tag{14}$$

Next, differentiating (14) and substituting the value of $\dot{\alpha}(t)$ from (13), we get

$$\upsilon(f_0(t+\alpha(t)))\cdot b(f_1 t) = \frac{1}{f_0}\frac{d\Delta\phi(t)}{dt} + \frac{f_1 - f_0}{f_0}. \tag{15}$$

Substituting the value of $\alpha(t)$ from (14) in (15), we obtain a phase difference equation suitable for further analysis

$$\frac{d\Delta\phi(t)}{dt} = -(f_1 - f_0) + f_0\upsilon(\Delta\phi(t) + \theta_1(t))\cdot b(\theta_1(t)). \tag{16}$$

In the phase difference equation (16), $\Delta\phi(t)$ is phase difference between the oscillator and the perturbation signal. For the injection locking analysis, perturbation signal is a weak periodic injection signal with frequency close to the oscillator's natural frequency.

We now assume that in (16), $\theta_1(t)$ is "fast" varying and $\Delta\phi(t)$ "slowly" varying variable. This assumption is explained below. Under this assumption, we can average out the fast $\theta_1(t)$ within a period and retain the slow $\Delta\phi(t)$ variations in (16). Therefore, after averaging out the fast variations, (16) can be written as

$$\frac{d\Delta\phi(t)}{dt} = -(f_1 - f_0) + f_0 g(\Delta\phi(t)), \tag{17}$$

where $g(\Delta\phi(t))$ is

$$\begin{aligned} g(\Delta\phi(t)) &= \frac{1}{\theta_{1_f} - 0}\int_0^{\theta_{1_f}} \upsilon(\Delta\phi(t) + \theta_1(t))\cdot b(\theta_1(t))\, d\theta_1(t), \quad \text{and} \\ \theta_{1_f} &= \theta_1\left(t = \frac{1}{f_1}\right) = 1. \end{aligned} \tag{18}$$

This is the Generalized Adler's equation valid for any oscillator as compared to the original Adler's equation (8), which is only applicable to LC-tank like oscillators. Note that, this is of same form as original Adler's equation but with different $g(\Delta\phi(t))$ depending upon the oscillator type, and the injection signal.

Under lock conditions, the phase difference between oscillator and injected signal becomes constant, that is $d\Delta\phi(t)/dt = 0$ or $\Delta\phi(t) = \Delta\phi_0$ and hence

$$\begin{aligned} f_1 - f_0 &= f_0 g(\Delta\phi_0), \quad \text{or} \\ \Delta f_0 &= f_0 g(\Delta\phi_0). \end{aligned} \tag{19}$$

Let f_L be the locking range of the oscillator. The maximum value of $g(\cdot)$ gives the locking range of the oscillator about f_0 as

$$\begin{aligned} |\Delta f_0|_{\max} &= f_0 \left[g(\Delta\phi(t))\right]_{\max}, \quad \text{and} \\ f_L &= 2|\Delta f_0|_{\max}. \end{aligned} \tag{20}$$

The locking range, f_L, is typically much smaller than f_1 or f_0. If the oscillator is not locked $\Delta\dot{\phi}(t) \neq 0$, and the maximum value of *RHS* in (17) is $-(f_1 - f_0) + f_L/2 \ll f_1$. Thus, we have

$$\begin{aligned} \left(\frac{d\Delta\phi(t)}{dt}\right)_{\max} &= -(f_1 - f_0) + f_L/2, \quad \text{and} \\ \frac{d\phi_1(t)}{dt} &= f_1. \end{aligned} \tag{21}$$

Because f_1 is close to free running frequency f_0 of the oscillator and f_L is small, the relationship

$$\left(\frac{d\Delta\phi(t)}{dt}\right)_{\max} \ll \frac{d\phi_1(t)}{dt} \tag{22}$$

holds. Hence, our assumption that $\phi_1(t)$ is fast varying and $\Delta\phi(t)$ is slowly varying is justified.

Now, we apply the Gen-Adler's equation on two oscillators, the ring oscillator and the Hodgkin-Huxley neuronal oscillator perturbed by a sinusoidal injection signal

$$b(\theta_1(t)) = I_i \sin(2\pi\theta_1(t)) \quad \text{if } 0 \leq \theta_1 < 1, \tag{23}$$

where $\phi_1(t) = f_1 t$.

3.3 Injection Locking Range of Ring Oscillator

In this subsection, we derive Gen-Adler's equation for the ring oscillator circuit shown in Fig. 1 with $R = 1$ kΩ, $C = 2$ pF, and $G_m = -50$. The DAEs of the ring oscillator circuit are given by (3). We first calculate the PPV waveforms of the ring oscillator using the Malkin's theorem, and are as shown in the Fig. 3a. Gen-Adler's equation for the ring oscillator is then obtained by evaluating $g(\Delta\phi(t))$ from (18), and is shown in Fig. 5. The lock range can be easily obtained by finding the maximum of $g(\Delta\phi(t))$. The injection current amplitude was taken to be $I_i = 0.1$ mA and the frequency, $f_1 = 1.01 f_0$, where f_0 is free running frequency of the ring oscillator.

The steady state phase difference, $\Delta\phi_0$, between the oscillator and the injected signal is given by "stable" solution of (19). For $\Delta f_0/f_0 = 0.01$, Fig. 5 shows the plot of (19). It can be clearly seen from Fig. 5 that oscillator will lock to the injection signal only if $\Delta f_0/f_0$ line intersects $g(\Delta\phi(t))$, that is (19) has a valid solution. The line $\Delta f_0/f_0$ intersects $g(\Delta\phi(t))$ twice in a period. The two intersection points are marked as U (unstable) and S (stable). When $\Delta\phi(t)$ is perturbed slightly from the unstable point, it will move away from it towards a stable point, depending upon the sign of $d\Delta\phi(t)/dt$ as shown in Fig. 5. For given $g(\Delta\phi(t))$, the locking range is the maximum range of injection signal frequency, f_1, yielding a valid solution of (19).

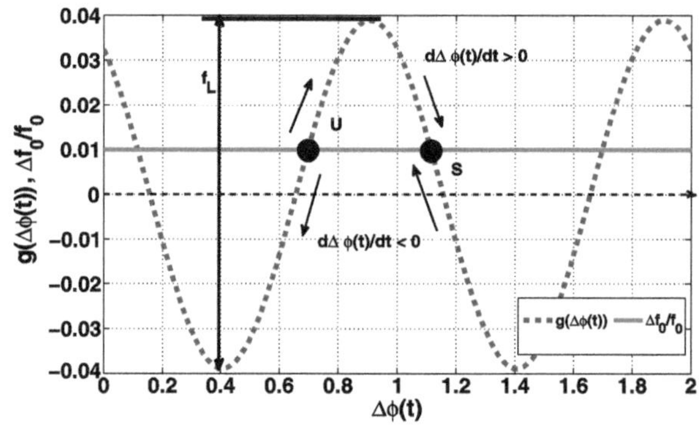

Fig. 5 Graphical solution of injection locking range in the ring oscillators with sinusoidal injection signal

In [14], analytical equations of $g(\Delta\phi)$ for ideal ring oscillator ($G_m = -\infty$) are obtained as

$$\boxed{\begin{aligned} g(\Delta\phi(t)) = \frac{1}{\sqrt{4\pi^2 + K_0^2}} \frac{RI_i}{A} \sin(2\pi\Delta\phi(t) + \zeta) \\ \times \left[K_1(e^{K_0/2} + 1) - K_2(e^{K_0} + e^{K_0/2})\right] \end{aligned}} \tag{24}$$

where,

$$\sin(\zeta) = \frac{2\pi}{\sqrt{4\pi^2 + K_0^2}}, \qquad K_0 = 2.887, \qquad K_1 = \frac{1}{\sqrt{5}}, \qquad K_2 = \left(\frac{2}{\sqrt{5}} - 1\right). \tag{25}$$

The phase lock would occur, when $\frac{d\Delta\phi(t)}{dt}$ is zero, and the locking range is given at the maximum value of $g(\Delta\phi(t))$. In this case, the maximum value of $g(\Delta\phi(t))$ occurs when $\sin(2\pi\Delta\phi(t)+\zeta)=1$, therefore

$$
\begin{aligned}
|\Delta f_0|_{\max} &= \frac{f_0}{\sqrt{4\pi^2+K_0^2}}\frac{RI_i}{A}\left[K_1(e^{K_0/2}+1)-K_2(e^{K_0}+e^{K_0/2})\right] \\
&= 0.6773 f_0 \frac{RI_i}{A},
\end{aligned}
\tag{26}
$$

and the locking range is $f_L = 2|\Delta f_0|_{\max}$.

3.4 Injection Locking Range of Hodgkin-Huxley Neuron

In this subsection, we apply the Gen-Adler's methodology to analyze injection locking in Hodgkin-Huxley neuron in oscillatory mode [8, 15]. The equations of the Hodgkin-Huxley model are

$$
\begin{aligned}
C\frac{dV}{dt} &= \overbrace{g_K n^4(e_K - V) + g_{\mathrm{Na}} m^3 h(e_{\mathrm{Na}} - V) + g_L(e_L - V)}^{\text{Ionic currents, } I_{\text{ionic}}} + I_{\mathrm{dc}} + I_{\mathrm{inj}}(t), \\
\frac{dn}{dt} &= a_n(1-n) - b_n n, \\
\frac{dm}{dt} &= a_m(1-m) - b_m m, \\
\frac{dh}{dt} &= a_h(1-h) - b_h h, \\
a_m &= \frac{0.1(V+40)}{(1-e^{-(V+40)/10)}}, \qquad b_m = 4e^{-(V+65)/18}, \\
a_n &= \frac{0.01(V+55)}{(1-e^{-(V+55)/10)}}, \qquad b_n = 0.125e^{-(V+65)/80}, \\
a_h &= 0.07e^{-(V+65)/20}, \quad \text{and} \quad b_h = \frac{1}{(1+e^{-(V+35)/10)})}
\end{aligned}
\tag{27}
$$

where, $C = 1\ \mu\mathrm{F/cm}^2$; $I_{\mathrm{dc}} = 14\ \mu\mathrm{A}$; $g_K = 36\ \mathrm{mS/cm}^2$; $g_{\mathrm{Na}} = 120\ \mathrm{mS/cm}^2$; $g_L = 0.3\ \mathrm{mS/cm}^2$; $e_K = -72$ mV; $e_{\mathrm{Na}} = 55$ mV; and $e_L = -49$ mV. The first state variable, $V(t)$, is known as the action potential of a neuron. We obtain the injection locking range for Hodgkin-Huxley neuron when perturbed by a sinusoidal injection signal, $I_{\mathrm{inj}}(t)$. First, we calculate the PPV waveforms of the neuron using the Malkin's theorem. The PPV waveform for action potential is shown in the Fig. 6a. Gen-Adler's equation for the Hodgkin-Huxley neuron is then obtained by evaluating $g(\Delta\phi(t))$ from (18), and is shown in Fig. 6b. The injection current amplitude was taken to be $I_i = 0.2\ \mu\mathrm{A}$ and the frequency, $f_1 = 1.01 f_0$, where $f_0 = 83.2$ Hz is free running frequency of the neuronal oscillator for Fig. 6b. The lock range can be easily obtained by finding the maximum of $g(\Delta\phi(t))$. Furthermore, the $g(\Delta\phi(t))$ plot enables us to easily predict the results instantly. For example, if $f_1 = 1.01 f_0$, there exist a solution of (19) and the neuronal oscillator would lock to the external injection signal. However, if $f_1 = 1.02 f_0$, we can easily see that the injection locking would not occur. This can be confirmed by running full transient simulation of the system as shown in Fig. 7. In Fig. 7a, the injection signal is $b(t) = 0.2\sin(2\pi 1.01 f_0 t)$ μA, and as it can be seen the oscillator is injection locked to the $b(t)$. In Fig. 7b, the injection signal is $b(t) = 0.2\sin(2\pi 1.02 f_0 t)$ μA and the oscillator is not injection locked to the $b(t)$.

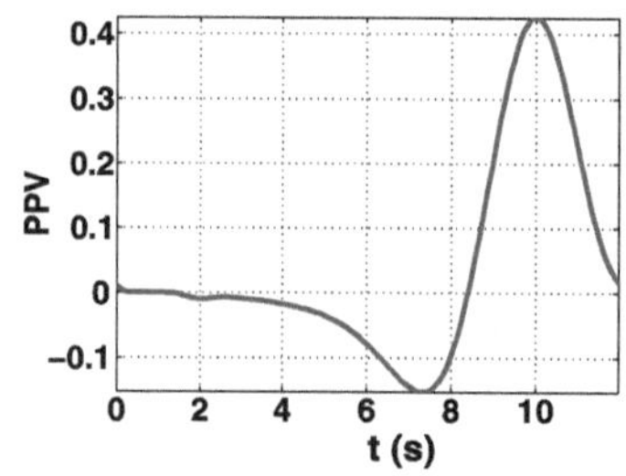

(a) PPV for action potential.

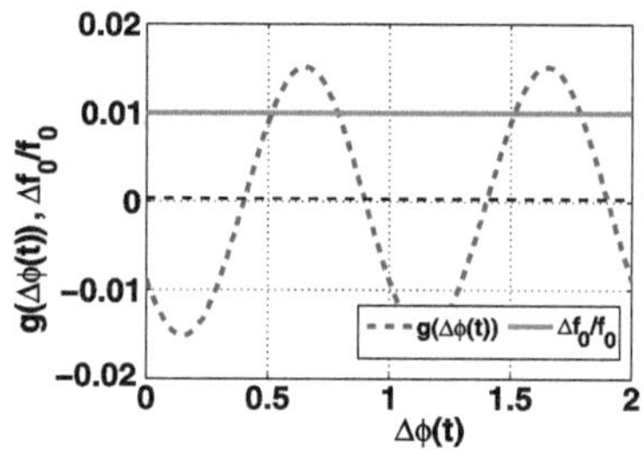

(b) $g(\Delta\phi(t))$ of the neuron.

Fig. 6 PPV for action potential and $g(\Delta\phi(t))$ of Hodgkin-Huxley neuron

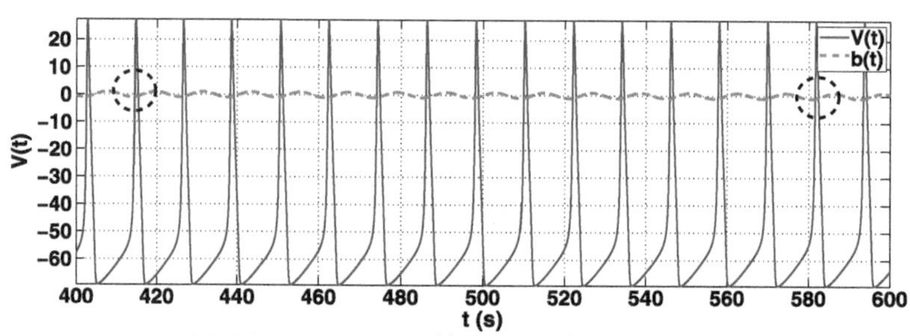

(a) Injection locked, $b(t) = 0.2\sin(2\pi 1.01 f_0 t)$ μA.

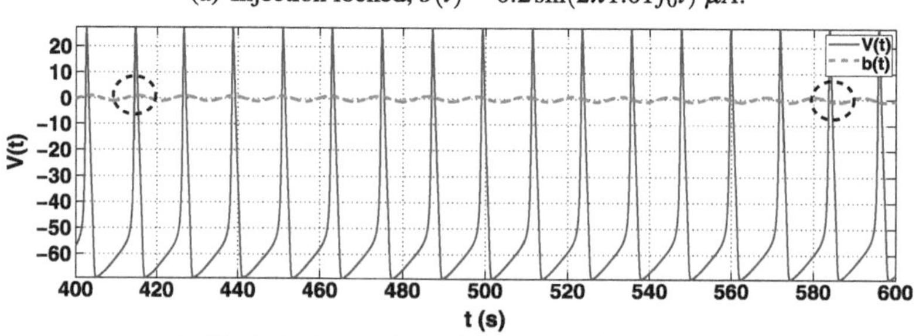

(b) Not injection locked, $b(t) = 0.2\sin(2\pi 1.02 f_0 t)$ μA.

Fig. 7 Validation for Gen-Adler's predictions. (Amplitude of $b(t)$ normalized to unity for plotting)

4 Coupled Oscillator Network Simulation

In this section, we use the PPV phase macromodel for efficient simulation of the weakly coupled CONs. Based on the PPV phase macromodel, we provide recipes of performing efficient transient and synchronized steady state simulation.

4.1 Coupled Oscillator Network Transient Simulation

Consider a system composited from N weakly coupled oscillators of the form (2). The coupled oscillator system is defined by the nN DAEs

$$\frac{d\mathbf{q}_i(\mathbf{x}_i(t))}{dt} = \mathbf{r}_i(\mathbf{x}_i(t)) + \mathbf{g}_i(\mathbf{X}(t)) \quad \text{for } i = 1,\ldots,N, \tag{28}$$

where $\mathbf{X}(t) : \mathbb{R} \mapsto \mathbb{R}^{nN}$ is $[\mathbf{x}_1(t),\ldots,\mathbf{x}_i(t),\ldots,\mathbf{x}_N(t)]^T$. The coupling between the oscillators is denoted by the function $\mathbf{g}_i(.) : \mathbb{R}^{nN} \mapsto \mathbb{R}^n$. Assume that when the oscillators are uncoupled and the external perturbation is zero they have the free running frequencies as: $\{f_1,\ldots,f_i,\ldots,f_N\}$. Thus, the phase of the ith oscillator is given as $\theta_i(t) = f_i(t + \alpha_i(t))$. Also, for the ith oscillator, rewriting (4) in terms of its phase we obtain

$$\begin{aligned}\mathbf{x}_i(t) &= \chi_{\mathbf{i}}(f(t + \alpha_i(t))) \\ &= \chi_{\mathbf{i}}(\theta_i(t)),\end{aligned} \tag{29}$$

where $\chi(.)$ is 1-periodic form of the steady state such that $\mathbf{x}_{ss}(t) = \chi(ft)$. Under the assumption of weak coupling, we can now use the phase macromodel (12) of individual oscillators to obtain a phase domain description of CON given by

$$\frac{d\theta_i(t)}{dt} = f_i + f_i \left[\upsilon_i(\theta_i(t)) \cdot \overbrace{\mathbf{g}_i(\chi(\theta(t)))}^{\text{coupling terms}} \right] \quad \text{for } i = 1,\ldots,N, \tag{30}$$

where $\chi(\cdot) = [\chi_1(\cdot),\ldots,\chi_i(\cdot),\ldots,\chi_N(\cdot)]^T$ and $\theta(t) = [\theta_1(t),\ldots,\theta_i(t),\ldots,\theta_N(t)]^T$. Note that the equation system (30) is of size $\mathbb{R}^N$ as compared to $\mathbb{R}^{nN}$ for the full system description as in (28). This reduced order phase domain model can be used for efficient coupled oscillator simulation. The solution of (30) can be used to compute CON state variables as

$$\mathbf{x}_i(t) = \chi_i(\theta_i(t)). \tag{31}$$

We summarize the overall procedure of CON transient simulation in Procedure 1. Further details and examples on transient simulation using the PPV macromodel can be found in [17].

Procedure 1 Transient simulation using the PPV phase macromodel

1: Compute the steady state solution of each oscillator using any periodic steady state analyzes such as Harmonic Balance (HB), Shooting or FDTD techniques [16].
2: Apply Malkin's theorem and compute PPVs of oscillators in the CON using (6).
3: Assemble the phase domain description of the CON as in (30) and solve the system of equation to obtain $\theta(t)$.
4: Finally obtain the state variables of the CON using (31).

4.2 *Finding Oscillator Phases in the Synchronized State of a CON Numerically*

Assume that the frequencies of the oscillators and the inter-oscillator coupling in the CON are appropriate for the phenomenon of synchronization to take place. When a CON is synchronized, all the oscillators change their frequency to achieve a common frequency, $f_c = 1/T_c$. Depending on the coupling, the common frequency may not correspond to the frequency of a particular oscillator in the network or not even an average of the frequencies of all oscillators in the network. In this subsection, a numerical method for computing the synchronized frequency, and the oscillator phases in a synchronized CON is presented.

In [18], a modified HB method called PPV-HB was proposed for oscillators and PLL systems described using the PPV phase macromodels. HB is directly applicable to autonomous oscillator differential equations because the state variables are periodic. For efficient oscillator related simulations the phase macromodelling techniques as discussed in Sect. 2 are employed. This effectively reduces the task to solving (12), a reduced order model, instead of (1). However, the solution of the phase equation of an oscillator, (12), features an unboundedly increasing phase term, and consequently standard HB cannot be applied directly. The PPV-HB technique decomposes the phase term into an unboundedly increasing part and a bounded periodic part. This decomposition enables use of efficient steady state simulation technique (HB method in the PPV-HB) for oscillator phase macromodel simulations. Adapting approaches from the PPV-HB technique [18], we hypothesize that absent any external perturbation, the phases of oscillators in the synchronized state follows the form

$$\theta_i^{\text{sync}}(t) = f_i(t + \alpha_i(t)) = f_c t + p_i(f_c t) \quad \forall i \in \{1, \ldots, N\}, \tag{32}$$

where the $p_i(\cdot)$'s are 1-periodic functions and f_c is synchronization frequency (both to be determined using FDTD simulations as described later). Thus, the phase consists of an unboundedly increasing phase term and a periodic function within each period of the synchronized state. In the absence of perturbation, a CON can be described by (30). Assume that the ith oscillators have a free running frequency f_i then the phase of the ith oscillator would be given by (32). The phases of the oscillators in a CON feature unboundedly increasing phase terms. Hence, we cannot apply steady state methods directly on (30) for synchronized state phase computation. Using the decomposition of phase as in PPV-HB, we propose a time-domain collocation or

FDTD technique for an efficient computation of steady state phase variations in a synchronized CON. Differentiating (32) gives

$$\frac{d\theta_i^{\text{sync}}(t)}{dt} = f_c + f_c \frac{dp_i(f_c t)}{dt} \quad \forall i \in \{1,\ldots,N\}, \tag{33}$$

where f_c is the common frequency of the oscillators after synchronization, and $p_i(\cdot)$s are 1-periodic functions, both of which are to be determined using numerical methods. Plugging (33) into (30), we obtain

$$\begin{aligned} f_c + f_c \frac{dp_i(f_c t)}{dt} &= f_i + f_i \left[\upsilon_i(\theta_i(t)) \cdot \mathbf{g}_i(\chi(\theta(t))) \right], \quad \text{or} \\ \frac{dp_i(f_c t)}{dt} &= \frac{f_i - f_c}{f_c} + \frac{f_i}{f_c} \left[\upsilon_i(\theta_i(t)) \cdot \mathbf{g}_i(\chi(\theta(t))) \right] \end{aligned} \tag{34}$$

for $i = 1,\ldots,N$. Since $\theta_i(t) = f_c t + p_i(f_c t)$, we have

$$\frac{dp_i(f_c t)}{dt} = \frac{f_i - f_c}{f_c} + \frac{f_i}{f_c} \left[\upsilon_i(f_c t + p_i(f_c t)) \cdot \mathbf{g}_i(\chi(f_c t + \mathbf{p}(f_c t))) \right], \tag{35}$$

where $\mathbf{p}(f_c t) = [p_1(f_c t),\ldots,p_N(f_c t)]^T$. Rewriting (35) in terms of the scaled time $t_s = f_c t$, we obtain

$$\frac{dp_i(t_s)}{dt_s} = \frac{f_i - f_c}{f_c} + \frac{f_i}{f_c} \left[\upsilon_i(t_s + p_i(t_s)) \cdot \mathbf{g}_i(\chi(t_s + \mathbf{p}(t_s))) \right] \tag{36}$$

for $i = 1,\ldots,N$. Assuming that the coupling function $\mathbf{g}_i(\cdot)$ is memoryless, $\mathbf{g}_i(\chi(t_s + \mathbf{p}(t_s)))$ will be periodic because $\chi(\cdot)$ is 1-periodic. Thus, in the above equation both sides are periodic since $p_i(\cdot)$ and $\upsilon_i(\cdot)$ are also periodic. Hence, we can apply steady state simulation methods for computation of the $p_i(\cdot)$s. We now formulate a time-collocation or FDTD technique for the computation of $p_i(\cdot)$s and hence the synchronized state phases, $\theta_i^{\text{sync}}(t)$. Equation (36) can be written in compact form as

$$\frac{d\mathbf{p}(t_s)}{dt_s} - \mathbf{q}(\mathbf{p}(t_s), t_s, f_c) = 0, \tag{37}$$

where

$$\begin{aligned} \mathbf{p}(t_s) &= [p_1(t_s),\ldots,p_N(t_s)]^T, \\ \mathbf{q}(\mathbf{p}(t_s),t_s) &= [q_1(\mathbf{p}(t_s),t_s),\ldots,q_N(\mathbf{p}(t_s),t_s)]^T, \quad \text{and} \\ q_i(\mathbf{p}(t_s),t_s,f_c) &= \frac{f_i - f_c}{f_c} + \frac{f_i}{f_c} \left[\upsilon_i(t_s + p_i(t_s)) \cdot \mathbf{g}_i(\chi(t_s + \mathbf{p}(t_s))) \right]. \end{aligned}$$

For numerical FDTD techniques, time is discretized into m steps, $\{t_0,\ldots,t_i,\ldots,t_m\}$, and the differential operator is replaced by a finite-difference approximation. The periodicity can be enforced by the condition $\mathbf{p}(t_m) = \mathbf{p}(t_0)$. Taking the uniform time-step h for time discretization such that $t_i = (i-1)h$, $i \in \{0,\ldots,m-1\}$ and the Backward Euler (for example) difference scheme [19] for the differential operator in (37) we have

$$\begin{aligned}\frac{\mathbf{p}(t_0)-\mathbf{p}(t_{m-1})}{h}-\mathbf{q}(\mathbf{p}(t_0),t_0,f_c)&=0,\\ \frac{\mathbf{p}(t_1)-\mathbf{p}(t_0)}{h}-\mathbf{q}(\mathbf{p}(t_1),t_1,f_c)&=0,\\ &\vdots\\ \frac{\mathbf{p}(t_{m-1})-\mathbf{p}(t_{m-2})}{h}-\mathbf{q}(\mathbf{p}(t_{m-1}),t_{m-1},f_c)&=0.\end{aligned} \tag{38}$$

The unknowns of the above finite difference equation are

$$\mathbf{X}_{\text{fd}}=[\mathbf{p}(t_0),\mathbf{p}(t_1),\ldots,\mathbf{p}(t_{m-1}),f_c]^T\in\mathbb{R}^{mN+1}. \tag{39}$$

Hence, the above nonlinear equations have only mN equations but $mN+1$ unknowns—N components of $\mathbf{p}(\cdot)$ at m time points and an extra unknown for the synchronized state CON frequency, f_c. This problem of an extra unknown is due to the lack of a "time-reference", or in other words the fact that if $\mathbf{p}(t_s)$ is a periodic solution, then $\mathbf{p}(t_s+\tau),\forall\tau$ is also a valid periodic solution of the above equations. This problem is solved by imposing a phase condition as done in the single oscillator PSS techniques [16]. A possible phase condition is to fix one of the entries of $\mathbf{p}(\cdot)$ at some fixed time. For example, fixing the value of the first entry of $\mathbf{p}(\cdot)$ at t_0 as p_{10} we have

$$p_1(t_0)=p_{10}. \tag{40}$$

Therefore, time-collocation equations gets appended to

$$\begin{bmatrix}\frac{\mathbf{p}(t_0)-\mathbf{p}(t_{m-1})}{h}-\mathbf{q}(\mathbf{p}(t_0),t_0,f_c)\\ \frac{\mathbf{p}(t_1)-\mathbf{p}(t_0)}{h}-\mathbf{q}(\mathbf{p}(t_1),t_1,f_c)\\ \vdots\\ \frac{\mathbf{p}(t_{m-1})-\mathbf{p}(t_{m-2})}{h}-\mathbf{q}(\mathbf{p}(t_{m-1}),t_{m-1},f_c)\\ p_1(t_0)-p_{10}\end{bmatrix}_{(mN+1)\times 1}=0. \tag{41}$$

Let us denote the left hand side of (41) by $\mathbf{H}_{\text{fd}}:\underbrace{\mathbb{R}^N\times\cdots\times\mathbb{R}^N}_{\text{m}}\times\mathbb{R}\mapsto\mathbb{R}^{mN+1}$. Thus, we need to solve the nonlinear equation system

$$\begin{aligned}\mathbf{H}_{\text{fd}}(\mathbf{p}(t_0),\mathbf{p}(t_1),\ldots,\mathbf{p}(t_{m-1}),f_c)=0,\quad\text{or}\\ \mathbf{H}_{\text{fd}}(\mathbf{X}_{\text{fd}})=0,\end{aligned} \tag{42}$$

to compute the steady state phases of a CON. The above set of $mN+1$ nonlinear equations can be solved using the Newton-Raphson method [20]. The kth iteration of the Newton-Raphson method requires the Jacobian, $J_{\text{fd}}(\mathbf{X}_{\text{fd}})_{(mN+1)\times(mN+1)}$, of (42) as

$$J_{\text{fd}}(X_{\text{fd}}^k)[\mathbf{X}_{\text{fd}}^{k+1}-\mathbf{X}_{\text{fd}}^k]=-\mathbf{H}_{\text{fd}}(\mathbf{X}_{\text{fd}}^k). \tag{43}$$

The Jacobian, $J_{\text{fd}}(\mathbf{X}_{\text{fd}})$, is given by the block matrix

$$J_{\text{fd}}(\mathbf{X}_{\text{fd}}) = \frac{\partial \mathbf{H}_{\text{fd}}(\mathbf{X}_{\text{fd}})}{\partial \mathbf{X}_{\text{fd}}} = \left[\begin{array}{c|c} P_{mN\times mN} & Q_{mN\times 1} \\ \hline R_{1\times mN} & S_{1\times 1} \end{array}\right], \tag{44}$$

where

$$P = \begin{bmatrix} \frac{C_0}{h} - G_0 & 0 & \cdots & -\frac{C_{m-1}}{h} \\ -\frac{C_0}{h} & \frac{C_1}{h} - G_1 & \cdots & 0 \\ \vdots & \ddots & \ddots & \cdots \\ 0 & 0 & -\frac{C_{m-2}}{h} & \frac{C_{m-1}}{h} - G_{m-1} \end{bmatrix},$$

$$Q = \begin{bmatrix} -\frac{\partial \mathbf{q}(\mathbf{p}(t_0),t_0,f_c)}{\partial f_c} \\ -\frac{\partial \mathbf{q}(\mathbf{p}(t_1),t_1,f_c)}{\partial f_c} \\ \cdots \\ -\frac{\partial \mathbf{q}(\mathbf{p}(t_{m-1}),t_{m-1},f_c)}{\partial f_c} \end{bmatrix}, \qquad R = \begin{bmatrix} 1 \; 0 \cdots 0 \end{bmatrix}, \tag{45}$$

$$S = 0,$$
$$C_i = C(t_i)_{N\times N} = \text{Identity matrix}, \quad \text{and}$$
$$G_i = G(t_i)_{N\times N} = \left.\frac{\partial \mathbf{q}(\mathbf{p}(t_i),t_i,f_c)}{\partial \mathbf{p}(\mathbf{t_i})}\right|_{\mathbf{p}(t_i)}.$$

The convergence of the Newton-Raphson iteration (43) would give PSS of $\mathbf{p}(t_s)$ and synchronization frequency, f_c. Thus, the phases of the oscillators after synchronization can be stated as

$$\theta^{\text{sync}}(t) = \begin{bmatrix} \theta_1^{\text{sync}}(t) \\ \vdots \\ \theta_N^{\text{sync}}(t) \end{bmatrix}_{N\times 1} = \begin{bmatrix} f_c t + p_1(f_c t) \\ \vdots \\ f_c t + p_N(f_c t) \end{bmatrix}_{N\times 1}, \tag{46}$$

or equivalently the phase deviations in the synchronized steady state are given as

$$\alpha^{\text{sync}}(t) = \begin{bmatrix} \alpha_1^{\text{sync}}(t) \\ \vdots \\ \alpha_N^{\text{sync}}(t) \end{bmatrix}_{N\times 1} = \begin{bmatrix} \frac{f_c t + p_1(f_c t)}{f_1} - t \\ \vdots \\ \frac{f_c t + p_N(f_c t)}{f_2} - t \end{bmatrix}_{N\times 1}. \tag{47}$$

4.3 PPV based Simulation of Ring Oscillator and Neuronal Oscillator Network

In this section, we illustrate the synchronized state phase computation by FDTD technique developed in the previous section using two examples. The CONs are composited from 5 oscillators in both the cases and the connection between the oscillators is shown in Fig. 8b. For each of the two examples we assemble the phase domain description of the CON as in (30), and perform a transient simulation following Procedure 1. Next, we apply the FDTD technique developed in the last section for synchronized state calculations.

4.3.1 A $-G_m$ LC Oscillator Based CON

The first example is of a CON composed from $-G_m$ LC oscillator's. A $-G_m$ LC oscillator is shown in Fig. 8a. The differential equations of a $-G_m$ LC oscillator are [21]

$$\begin{aligned} C\frac{dv(t)}{dt} &= -f(v) - i(t) - \frac{v(t)}{R}, \\ L\frac{di(t)}{dt} &= v(t), \end{aligned} \tag{48}$$

where $f(v(t)) = \frac{1}{R}\tanh(-\frac{1.1}{R^2}v(t))$ represents the nonlinear negative resistance, $v(t)$ is the voltage across capacitor, and $i(t)$ is the current through inductor.

We connect the oscillators through a resistive coupling between the $v(t)$ state variable. The topology of the CON is shown in Fig. 8b. Each ball in Fig. 8b represents an oscillator and the lines correspond to the resistive coupling between $v(t)$ state variable of neighboring oscillators. Thus, the system of equations corresponding to this CON is

$$\begin{aligned} C_1\frac{dv_1(t)}{dt} &= -f(v_1) - i_1(t) - \frac{v_1(t)}{R} + \frac{v_5(t)+v_2(t)-2v_1(t)}{R_c}, \\ C_j\frac{dv_j(t)}{dt} &= -f(v_j) - i_j(t) - \frac{v_j(t)}{R} \\ &\quad + \frac{v_{j+1}(t)+v_{j-1}(t)-2v_j(t)}{R_c} \quad \text{for } j \in \{2,3,4\}, \\ C_5\frac{dv_5(t)}{dt} &= -f(v_5) - i_5(t) - \frac{v_1(t)}{R} + \frac{v_1(t)+v_4(t)-2v_5(t)}{R_c}, \quad \text{and} \\ L_j\frac{di_j(t)}{dt} &= v_j(t) \quad \text{for } j \in \{1,2,3,4,5\}, \end{aligned} \tag{49}$$

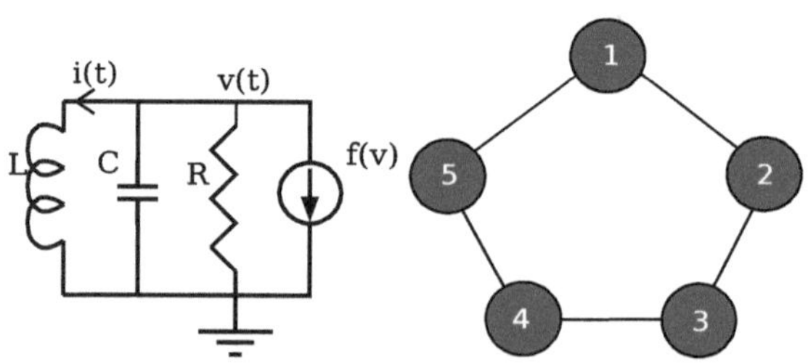

(a) A $-G_m$ LC oscillator. (b) CON topology.

Fig. 8 A LC oscillator and the CON topology

where R_c is the coupling resistance. We chose $R_c = 100\,\Omega$ and $R = 100\,\Omega$ for simulations. The L_is and C_is were chosen so as to have different frequencies across the CON with an initial frequency distribution as $f_1 = 1.1097$ GHz ; $f_2 = 1.0037 f_1$; $f_3 = 1.0228 f_1$; $f_4 = 1.0074 f_1$; and $f_5 = 0.98176 f_1$. For this initial frequency distribution and the coupling resistance the CON gets synchronized. First, the steady state solution of individual oscillator is obtained using the HB technique [16]. Next, the PPVs of oscillators is computed using Malkin's theorem (6). The 1-periodic form of steady state and PPV component corresponding to $v_1(t)$ in the CON equations (49) is shown in Fig. 9. Next, we compute the synchronized frequency and the steady state phase distribution across the CON using the FDTD technique proposed in 4.2. The solution of (42) yields $p_i(t)$s as shown in Fig. 10 and f_c (common synchronization frequency) as 1.1129 GHz. Based on our experiments, the Newton-Raphson on FDTD equation system (42) usually converges in 4–5 iterations. The results of the FDTD techniques gives excellent match with direct transient simulation results of (30) assembled for (49) and a comparison of phase deviation in synchronized state is shown in Fig. 11.

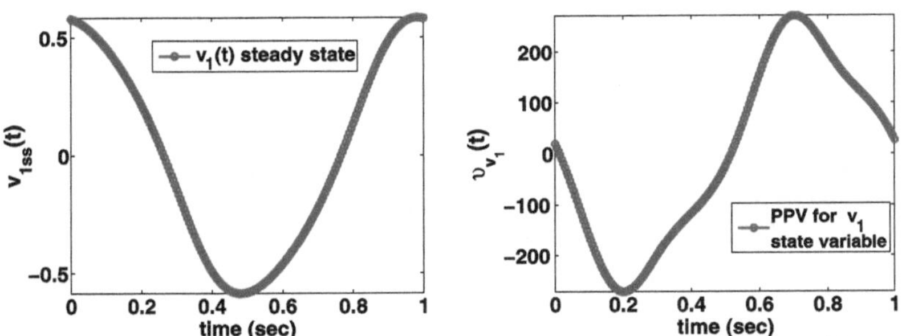

(a) 1-periodic $v_{1ss}(t)$. (b) 1-periodic PPV of $v_1(t)$.

Fig. 9 1-periodic form of steady state and a PPV component of first oscillator

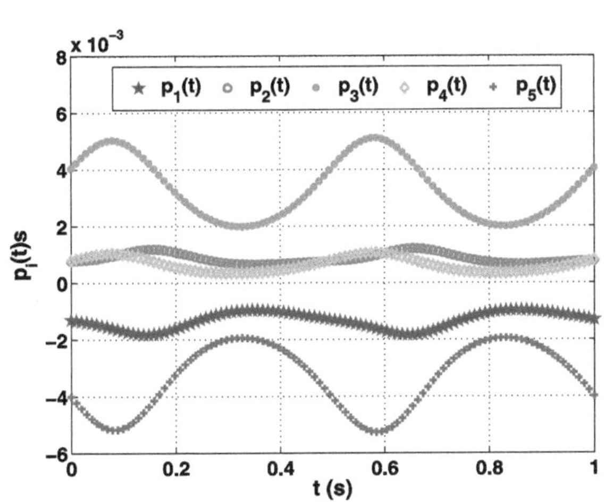

Fig. 10 The $p_i(t)$s of the oscillators in the LC CON computed using FDTD technique

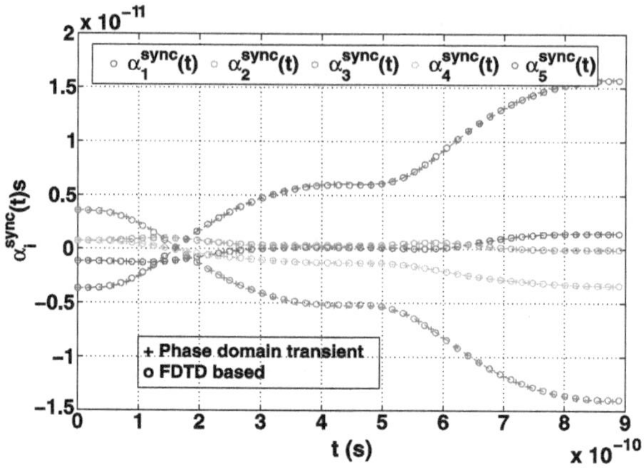

Fig. 11 Comparison of $\alpha_i^{\text{sync}}(t)$s of the oscillators in the LC CON computed using FDTD technique and phase domain transient simulation

4.3.2 A Hodgkin-Huxley Neuron Network

Our next example is of a simple neuronal oscillator network of the same size and topology as the previous network. We use the Hodgkin-Huxley neuron model in its oscillatory mode [8]. It has four state variables.

Let the action potential of ith neuron be given by $V_i(t)$. Using (27) we assemble the system of equations for this CON as

$$
\begin{aligned}
C\frac{dV_1}{dt} &= g_K n_1^4(e_K - V_1) + g_{\text{Na}} m_1^3 h_1 (e_{\text{Na}} - V_1) + g_L(e_L - V_1) + I_{\text{dc}_1} \\
&\quad + \frac{V_5(t) + V_2(t) - 2V_1(t)}{R_c}, \\
C\frac{dV_i}{dt} &= g_K n_i^4(e_K - V_i) + g_{\text{Na}} m_i^3 h_i (e_{\text{Na}} - V_i) + g_L(e_L - V_i) + I_{\text{dc}_i} \\
&\quad + \frac{V_{i+1}(t) + V_{i-1}(t) - 2V_i(t)}{R_c} \quad \text{for } i \in \{2,3,4\}, \\
C\frac{dV_5}{dt} &= g_K n_5^4(e_K - V_5) + g_{\text{Na}} m_5^3 h_5 (e_{\text{Na}} - V_5) + g_L(e_L - V_5) + I_{\text{dc}_5} \\
&\quad + \frac{V_1(t) + V_4(t) - 2V_5(t)}{R_c},
\end{aligned}
\tag{50}
$$

and

$$
\begin{aligned}
\frac{dn_i}{dt} &= a_{n_i}(1-n_i) - b_{n_i}n_i, \\
\frac{dm_i}{dt} &= a_{m_i}(1-m_i) - b_{m_i}m_i, \\
\frac{dh_i}{dt} &= a_{h_i}(1-h_i) - b_{h_i}h_i, \\
a_{m_i} &= \frac{0.1(V_i+40)}{(1-e^{-(V_i+40)/10})}, \qquad b_{m_i} = 4e^{-(V_i+65)/18}, \\
a_{n_i} &= \frac{0.01(V_i+55)}{(1-e^{-(V_i+55)/10})}, \qquad b_{n_i} = 0.125e^{-(V_i+65)/80}, \\
a_{h_i} &= 0.07e^{-(V_i+65)/20}, \quad \text{and} \quad b_{h_i} = \frac{1}{(1+e^{-(V_i+35)/10)})} \quad \text{for } i \in 1,2,3,4,5,
\end{aligned}
\tag{51}
$$

where $C = 1$ μF/cm^2; $g_K = 36$ mS/cm^2; $g_{\text{Na}} = 120$ mS/cm^2; $g_L = 0.3$ mS/cm^2; $e_K = -72$ mV; $e_{\text{Na}} = 55$ mV; and $e_L = -49$ mV. Next, we proceed in a similar manner to that of our previous example. The I_{dc_i} parameter of Hodgkin-Huxley oscillators in the CON were chosen so as to have an initial frequency distribution $f_1 = 83.2015$ Hz; $f_2 = 1.001f_1$; $f_3 = 1.005f_1$; $f_4 = 1.004f_1$; and $f_5 = 0.998f_1$. For this frequency distribution and the coupling resistance, $R_c = 100$, the CON gets synchronized. First, we compute the synchronized state phases and synchronization frequency using the FDTD technique. The synchronization frequency, f_c, was found to be 83.45 Hz. The $p_i(t)$s, and a comparison of $\alpha_i^{\text{sync}}(t)$s from FDTD technique against the transient simulation of phase domain description are shown in Figs. 12 and 13.

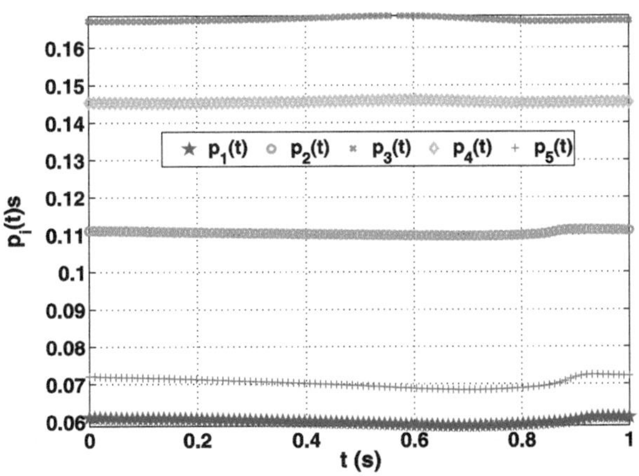

Fig. 12 The $p_i(t)$s of the oscillators in the neuron CON computed using FDTD technique

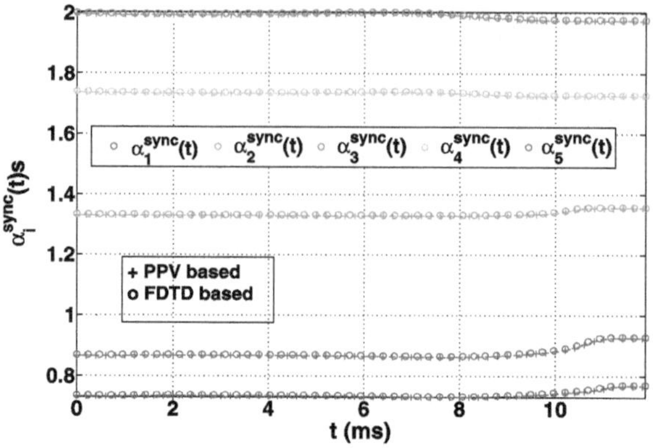

Fig. 13 The $\alpha_i^{\mathrm{sync}}(t)$s of the oscillators in the neuron CON computed using FDTD technique

5 Conclusions

In this chapter, we presented oscillator phase macromodel based injection locking analysis, and simulation techniques for coupled oscillator networks. The Generalized Adler's equation can be used for quick and insightful injection locking analysis of any oscillator as demonstrated in the examples. A recipe for the transient simulation, and a detailed formulation along with the numerical method for synchronized state calculation is presented. The FDTD steady state method can be used for efficient computation of steady state phase distribution and the synchronization frequency of the oscillators in a synchronized CON.

Acknowledgements We thank the Gigascale Systems Research Center which supported this work.

References

1. A.T. Winfree. *The Geometry of Biological Time*. Springer Verlag, 2001.
2. R. Adler. A study of locking phenomena in oscillators. *Proceedings of the IRE*, 34(6):351–357, 1946.
3. X. Lai and J. Roychowdhury. Fast simulation of large networks of nanotechnological and biochemical oscillators for investigating self-organization phenomena. In *Asia and South Pacific Conference on Design Automation, 2006*, page 6, 2006.
4. Davit Harutyunyan, Joost Rommes, Jan Ter Maten, and Wil Schilders. Simulation of mutually coupled oscillators using nonlinear phase macromodels. *Trans. Comp.-Aided Des. Integ. Cir. Sys.*, 28(10):1456–1466, 2009.
5. Y. Kuramoto. *Chemical Oscillations, Waves, and Turbulence*. Springer, Berlin, 1984.
6. A. Demir, A. Mehrotra, and J. Roychowdhury. Phase noise in oscillators: a unifying theory and numerical methods for characterization. *IEEE Transactions on Circuits and Systems I: Fundamental Theory and Applications*, 47(5):655–674, 2000.

7. A. Demir and J. Roychowdhury. A reliable and efficient procedure for oscillator PPV computation, with phase noise macromodeling applications. *IEEE Transactions on Computer-Aided Design of Integrated Circuits and Systems*, 22(2), 2003.
8. Eugene M. Izhikevich. *Dynamical Systems in Neuroscience: The Geometry of Excitability and Bursting*. Chapter 10. The MIT Press, 2007.
9. M. Farkas. *Periodic Motions*. Springer-Verlag, New York, 1994.
10. P. Kinget, R. Melville, D. Long, and V. Gopinathan. An injection-locking scheme for precision quadrature generation. *IEEE Journal of Solid-State Circuits*, 37(7):845, 2002.
11. C.J.M. Verhoeven. A high-frequency electronically tunable quadrature oscillator. *IEEE Journal of Solid-State Circuits*, 27(7):1097–1100, July 1992.
12. S. Kudszus, M. Neumann, T. Berceli, and W.H. Haydl. Fully integrated 94-ghz subharmonic injection-locked pll circuit. *IEEE Transactions on Microwave and Guided Wave Letters*, 10(2):70–72, February 2000.
13. B. Razavi. A study of injection pulling and locking in oscillators. *IEEE Journal of Solid-State Circuits*, 39(9), September 2004.
14. P. Bhansali and J. Roychowdhury. Gen-Adler: The generalized Adler's equation for injection locking analysis in oscillators. In *Design Automation Conference, 2009. ASP-DAC 2009. Asia and South Pacific*, pages 522–527, 2009.
15. A.L. Hodgkin and A.F. Huxley. A quantitative description of membrane current and its application to conduction and excitation in nerve. *The Journal of Physiology*, 117(4):500, 1952.
16. K.S. Kundert, J.K. White, and A. Sangiovanni-Vincentelli. *Steady-State Methods for Simulating Analog and Microwave Circuits*. Kluwer Academic Publishers, 1990.
17. Prateek Bhansali, Shweta Srivastava, Xiaolue Lai, and Jaijeet Roychowdhury. Comprehensive procedure for fast and accurate coupled oscillator network simulation. In *ICCAD '08: Proceedings of the 2008 IEEE/ACM International Conference on Computer-Aided Design*, pages 815–820, Piscataway, NJ, USA, 2008. IEEE Press.
18. Ting Mei and Jaijeet Roychowdhury. PPV-HB: Harmonic Balance for Oscillator/PLL Phase Macromodels. In *ICCAD '06: Proceedings of the 2006 IEEE/ACM International Conference on Computer-Aided Design*, pages 283–288, November 2006.
19. L.O. Chua and P.M. Lin. *Computer-Aided Analysis of Electronic Circuits: Algorithms and Computation Techniques*. Prentice-Hall, Englewood Cliffs, NJ, 1975.
20. W.H. Press, S.A. Teukolsky, W.T. Vetterling, and B.P. Flannery. *Numerical Recipes—The Art of Scientific Computing*. Cambridge University Press, 1989.
21. Xiaolue Lai and Jaijeet Roychowdhury. Analytical equation for predicting injection locking in LC and ring oscillator. In *Proceedings of the IEEE Custom Integrated Circuits Conference*, pages 461–464, September 2005.

Dynamic Stability of Static Memories: Concepts and Advanced Numerical Analysis Techniques

Peng Li, Wei Dong and Garng M. Huang

Abstract Static noise margins have been traditionally used to characterize the stability of static memories such as SRAMs. Continuing technology scaling has significantly shrunk the stability margins of static memories. Due to their inability in capturing nonlinear cell dynamics, static noise margins become increasingly inappropriate for state-of-the-art SRAMs with shrinking access time and/or advanced dynamic read-write-assist circuits. Based upon understandings derived from rigorous nonlinear system theory, we define new static memory dynamic noise margin concepts. These new metrics not only capture key nonlinear dynamical stability properties but also provide valuable design insights. Furthermore, we discuss advanced numerical simulation techniques that are appropriate for analyzing dynamic stability with good robustness and efficiency.

1 Introduction

Static memories such as static random access memories (SRAMs), latches and flip-flops provide indispensable on-chip data storage and occupy a significant part of chip area in many chip designs. Stability is one of the most desirable design properties for any static memory cell designs. It manifests itself as proper data retention in hold, and nondestructive access and successful overwrite in read and write, re-

Peng Li
Department of Electrical and Computer Engineering, Texas A&M University, College Station, TX 77843, USA, e-mail: pli@tamu.edu

Wei Dong
Texas Instruments, 12500 TI Boulevard, MS 8714, Dallas, TX 75243, USA, e-mail: weidong@ti.com

Garng M. Huang
Department of Electrical and Computer Engineering, Texas A&M University, College Station, TX 77843, USA, e-mail: ghuang@neo.tamu.edu

P. Li et al. (eds.), *Simulation and Verification of Electronic and Biological Systems*,
DOI 10.1007/978-94-007-0149-6_5,

spectively. The conventional design metrics used for quantifying stability are static noise margins (SNMs) [1, 2]. Such SNM metrics are increasingly inappropriate for highly scaled SRAM designs as sub-90nm technologies has significantly shrunk SRAM stability margins [7]. Although simple to obtain under the DC operating condition, static noise margins cannot capture the fundamental nonlinear dynamics upon which SRAM cells operate. Practically, when SNMs are used, assumptions have been made that read, write or noise injections last forever. Apparently, this is inconsistent with high-speed SRAM designs with short access time or advanced timing control circuits for read/write assists [7, 8]. To this end, properly defining and examining dynamic stability of SRAMs is highly desirable.

To properly characterize dynamic stability of SRAM operations, we first develop understandings on the basic nonlinear dynamics of SRAM cells using rigorous nonlinear system theory. The notion of stability boundary, or separatrix [3, 9] is employed as primary tool for characterizing SRAM dynamic stability. Using the concept of separatrix, new dynamic noise margins (DNMs) are defined [10]. These DNMs not only characterize dynamic stability in a way relevant to basic SRAM operations, but also provide useful linkage between stability and key timing control parameters. Broadly, conventional SNMs can be considered as *special* cases of the presented more general DNMs.

While brute-force simulation techniques exist for computing the proposed DNMs, efficient analysis entails an employment of nonlinear system principles and proper handling of intricate numerical characteristics. We present efficient system theoretically motivated CAD algorithms that can be applied with SPICE-level accuracy for dynamic stability analysis. In particular, for SRAM cells that are modeled as a low-dimensional dynamic system, we present a fast separatrix tracing algorithm using system theory and use it as a basis for computing dynamic noise margins [10]. We show such approach can be orders of magnitude faster than a brute-force approach, making it very appealing for driving Monte Carlo based statistical dynamic stability analysis. We further suggest advanced numerical techniques that can address the numerical stability issues in the separatrix tracing algorithm [6, 11]. For more general high-dimensional SRAM cell netlists (e.g. ones with extracted layout parasitics), we discuss a separatrix tangent based approach that is more suitable for dealing with higher dimensionality [12].

Throughout this chapter, we use the standard 6-T SRAM to demonstrate the presented techniques. However, with straightforward extensions, the same techniques can be applied to other types of SRAMs and static memories such as latches.

2 Static Noise Margins

Conventional static noise margins (SNMs) characterize a memory cell's noise immunity under the DC condition, i.e. with the injection of static noises. SNMs can be computed in several different but equivalent ways [1]. Among these, for instance, static noise margins in hold and read can be determined as shown in Fig. 1. In hold, the SNM is determined to be the side of the largest square that can be inscribed between the mirrored DC voltage transfer curves (VTC) of the cross-coupled inverters [1], where the VTC characterizes the input and output voltage correspondence of an inverter. The resulting SNM corresponds to the largest differential DC voltage noise that can be tolerated at the two storage nodes by the memory cell without losing the stored state. In short, hold SNM characterizes the largest tolerable DC voltage noise in standby.

Similarly, SNM in read can be also defined by inscribing the largest square between a pair of inverter VTCs. Note that during the read access, the two access transistors are turned on, which effectively alters the pull-up strength of the cross-coupled inverter pair. With an access transistor included in the pull-up, a modified VTC can be determined for each inverter (shown as the dotted Butterfly curve in Fig. 1). As can be seen in the figure, due to the alteration of the VTCs, a smaller space exists between the two Butterfly curves, leading to a reduced static noise margin compared with that of hold. Electrically, SNM in read characterizes the largest DC voltage perturbation that can be tolerated by the cell without producing a state flip, or a destructive read. During a write, one bit line is discharged and the other is pre-charged. The two inverter VTCs do not form an enclosed region. Note that writability corresponds to a successful flip of the SRAM cell state. As such, SNM in write can be found by inscribing the smallest square in between the two VTCs.

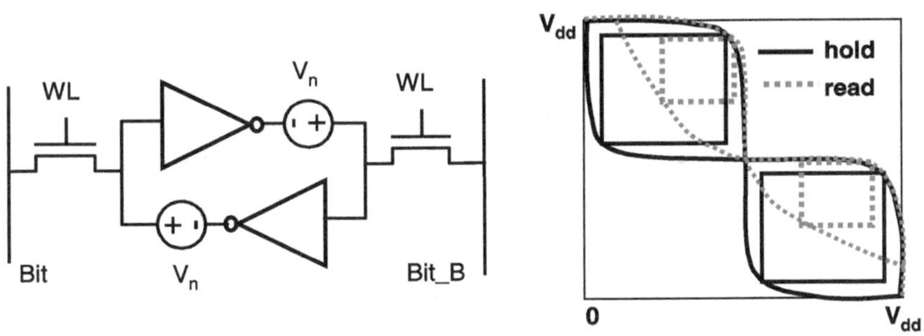

Fig. 1 Left: a standard 6-T SRAM cell; right: static noise margins in hold and read (from [10] ©[2008] IEEE)

Consistent with their static nature, the above static noise margins can be straightforwardly determined by performing DC circuit analysis. Clearly, such stability metrics do not capture any parasitic effects and hence cannot predict the cell stability in a dynamic fashion.

3 Dynamic Stability Boundaries of Bistable Systems

To characterize the stability of a memory cell from a dynamic perspective, we shall first examine the important concept of dynamic stability boundaries for a bistable system. The dynamic behavior of an SRAM cell can be described using the standard modified nodal analysis

$$f(x(t)) + \frac{d}{dt}q(x(t)) + u(t) = 0, \tag{1}$$

where $x(t) \in R^N$ is the vector of nodal voltages and branch currents, $u(t) \in R^N$ is the input, $f(\cdot)$ and $q(\cdot)$ are nonlinear functions describing static and dynamic nonlinearities.

We use the simple 2D case ($N = 2$) to simplify the visual presentation. The more general high-dimension case is discussed later in this chapter. Designed as a bistable nonlinear dynamical system, a memory cell has three equilibria, each of which is a DC solution to (1)

$$f(x_e) = 0. \tag{2}$$

Among the three equilibria, denote the two stable ones by x_e^s, corresponding to the *zero* and *one* states of the cell. The third one, denoted by x_e^u, is unstable and also referred to as the saddle.

To understand the stability of the system around an equilibrium, formally, the stable manifold of equilibrium x_e is defined as [3]

$$W^s(x_e) = \{x \in R^N \mid \lim_{t \to \infty} \phi(t,x) = x_e\}, \tag{3}$$

where $\phi(t,x)$ is the state trajectory that starts from the initial state x and finally converges to x_e. Intuitively, as depicted in Fig. 2, $W^s(x_e)$ defines the stability region of x_e in the state space. That is, any state perturbation that is confined within $W^s(x_e)$ will eventually diminish, and the state of the system will be attracted back to the equilibrium.

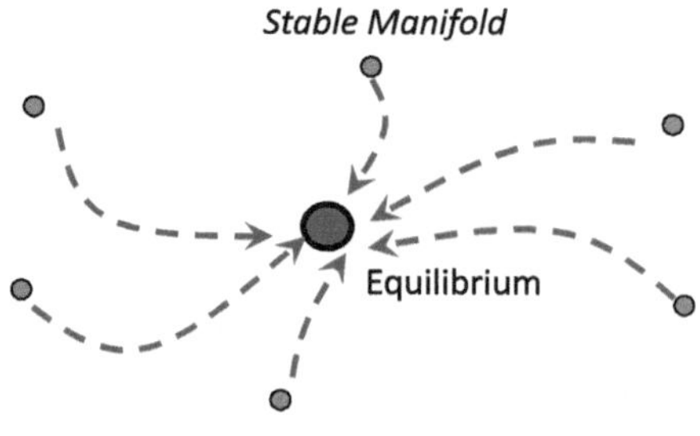

Fig. 2 An equilibrium and its stable manifold. The state trajectories that start in the stable manifold converge to the equilibrium

To apply the above concepts in a way relevant to the operations of an SRAM cell, it is instrumental to understand how the SRAM state may transit from one stable equilibrium to the other (i.e. a state flip) in hold, read or write. To illustrate, the phase portraits of two 6-T SRAM cells in the two-dimensional state space are shown in Fig. 3, where the first one is an ideal symmetric cell while the second one has device mismatch across the two cross-coupled inverters. The vector fields (time derivatives of the state variables) are shown by arrowed lines.

Any small perturbation around an stable equilibrium will eventually diminish and the transient SRAM state trajectory will be attracted towards to the corresponding stable equilibrium. Using the concepts defined above, the *stability region*, or *region of attraction*, of a stable equilibrium is defined to be its stable manifold. Furthermore, there exists a stability region for each stable equilibrium (i.e. the one or zero state) in the state space. The boundary between the two stability regions, or the *separatrix*, splits the entire state space. As discussed later, the separatrix is related to the stable manifold of the saddle. The separatrix plays a central role in characterizing the dynamic stability properties of the cell as the state moves from the one-state to the zero-stable or vice versa; a state flip will be generated if and only if the amplitude and duration of the disturbance to the cell are sufficiently high such that the state is pushed away from the initial stable equilibrium to cross the separatrix. It is important to note that the separatrix is a function of design or device parameters of the cell. For example, while the separatrix of the ideal symmetric cell is along the well expected 45° line in the state space, that of the latter is distorted by device parameter mismatch.

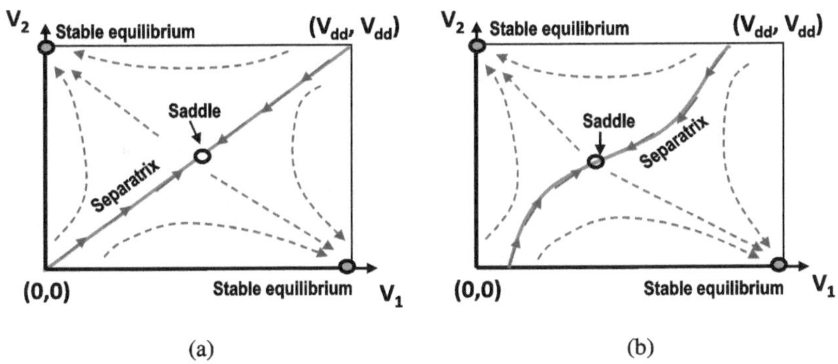

Fig. 3 Separatrices of two 6-T SRAM cells: (**a**) symmetric cell, and (**b**) unsymmetrical cell under parameter mismatch (modified from [10] ©[2008] IEEE)

4 Dynamic Noise Margins

The concept of stability boundary, or separatrix, facilitates the understanding of dynamic properties of a memory cell. Using it, new dynamic noise margins (DNMs) are defined for read, write and hold.

4.1 Dynamic Read Noise Margin (DRNM)

To properly define a dynamic noise margin for read operations, we shall consider the process of accessing an SRAM cell, as illustrated in Fig. 4a. Before the read process starts, the bit and bit-bar lines are fully charged. In addition to the cell being read, other unselected cells in the same column may draw leakage currents from either the bit or bit-bar lines depending on the stored value. These leakage contributions may be properly modeled to capture inter-cell interactions. The maintaining of dynamic stability corresponds to a successful nondestructive read during which the original state is retained while the cell is being accessed with a finite duration.

As such, it is of no surprise that DRNM is defined with respect to a given word-line pulse width T_R for read accesses, and based upon the separatrix. It is important to note, however, while defining the dynamic noise margin for read, the separatrix of the cell in hold (i.e. when the two access transistors are off) shall be used. As the read access proceeds (Fig. 4a), turning on the access transistors may push the state away from the initially stored *zero* or *one* state. If the state trajectory never crosses the separatrix in hold, the cell is statically stable and the time it takes for the transient trajectory to reach the separatrix is set to be $T_{\mathrm{across}} = \infty$; otherwise, the cell

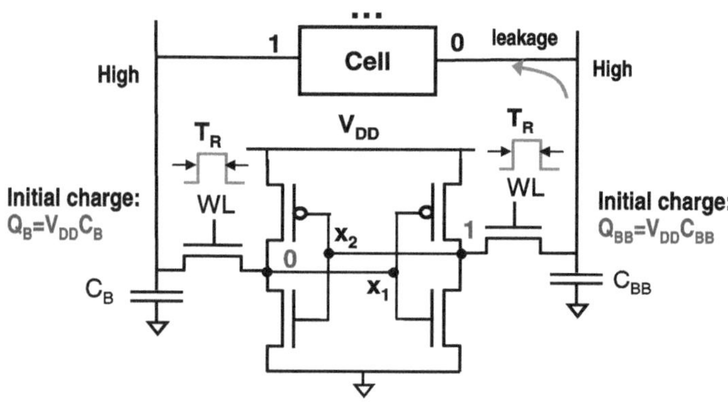

(a)

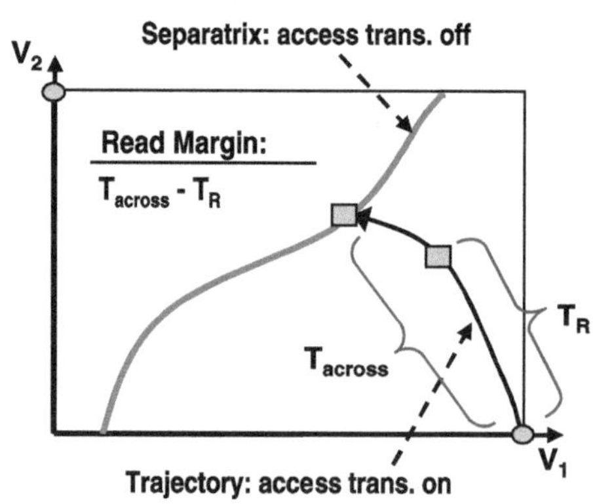

(b)

Fig. 4 Dynamic read noise margin: (**a**) read access of a 6-T SRAM cell, and (**b**) definition of DRNM (from [10] ©[2008] IEEE)

is statically unstable and T_{across} is finite. As shown in Fig. 4b, we have the following DRNM definition:

Definition 1 (Dynamic Read Noise Margin).

$$T_{\text{DRNM}} = \begin{cases} T_{\text{across}} - T_R, & \text{if statically unstable} \\ \infty, & \text{if statically stable} \end{cases} \tag{4}$$

This is an appropriate dynamic stability metric since a destructive read may be resulted if the cell state is pushed away from its initial value and goes across the separatrix (in hold) during the read access. At the end of the read process, the wordline goes off, the cell effectively returns to hold. Hence, if the state trajectory has already reached the other side of the separatrix, a state flip will incur.

It is not difficult to see that the DRNM defined as such specifies the amount of read access time margin before read instability takes place, hence dynamically characterizing the stability of the cell in read operations. More specifically, when $T_{\text{across}} > T_R$, there exists a positive, or potentially infinite, margin; when $T_{\text{across}} = T_R$, the cell is on the verge of read instability; when $T_{\text{across}} < T_R$, state-flip happens and the cell loses read stability. Obviously, DRNM directly relates to T_R, the read access time, which is further constrained by the minimum amount time required to sense the stored state across the bit and bit-bar lines. The two conflicting requirements on T_R must be simultaneously satisfied for a working SRAM cell design.

4.2 Dynamic Write Noise Margin (DWNM)

With write stability or writability, one is concerned of the ability of overwriting the state of a memory cell, which may involve forcing a state flip. While with this difference, dynamic write noise margin (DWNM) can be defined in a way analogous to that of DRNM. As shown in Fig. 5a, before a write access starts, one of the bit and bit-bar lines is discharged. Then, keeping the access transistors on may or may not push the cell state away the initial value to cross the separatrix in hold. In the former case, the cell is statically unstable in terms of writes and we set the separatrix crossing time to be $T_{\text{across}} = \infty$, signifying a static write failure. Otherwise, T_{across} is finite. With respect to a given wordline pulse width (write access time) T_W the DWNM is defined as (Fig. 5b)

Definition 2 (Dynamic Write Noise Margin).

$$T_{\text{DWNM}} = \begin{cases} T_W - T_{\text{across}}, & \text{if statically stable} \\ -\infty, & \text{if statically unstable} \end{cases} \tag{5}$$

Note that in the above equation $-\infty$ is used to denote a static write failure. As before, T_{DWNM} is a function of access time T_W. For statically stable cells, T_{DWNM} increases at the expenses of a longer access latency. In practice, dynamic noise margin and access latency shall be properly balanced to yield an optimized design.

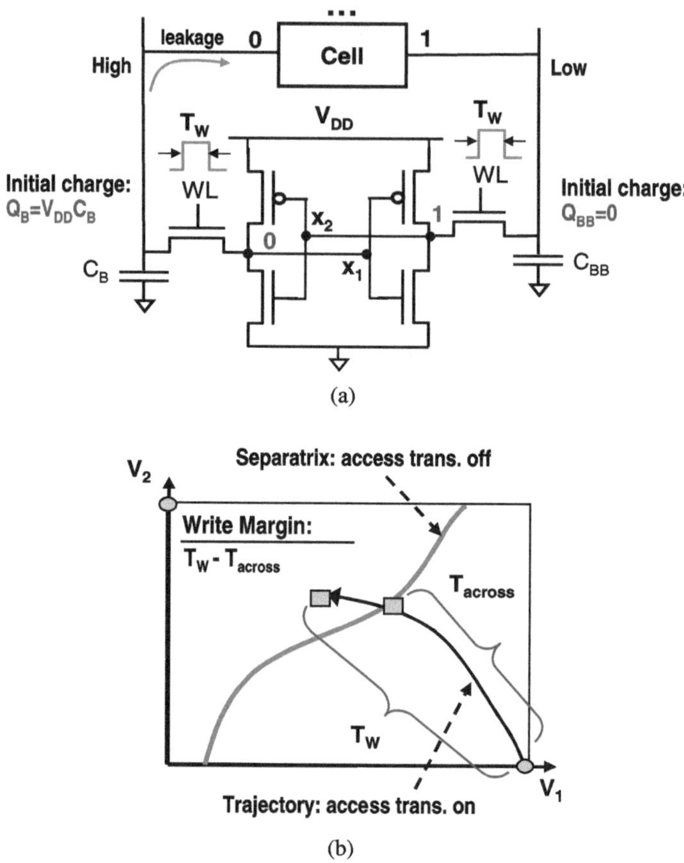

Fig. 5 Dynamic write noise margin: (**a**) write access of a 6-T SRAM cell, and (**b**) definition of DWNM (from [10] ©[2008] IEEE)

4.3 Dynamic Hold Noise Margin (DHNM)

Under the same spirit, the presented dynamic stability in hold (DHNM) characterizes the data retention property of the cell in standby. Different from traditional static noise margins [1], here the stability property is reflected by examining the cell's immunity to injected current disturbances, which more physically reflects the nonlinear dynamic nature of the cell. Additionally, current injections with different magnitudes and durations can be considered. For a given current noise amplitude, as shown in Fig. 6, the time it takes for the current noise to push the state trajectory across the separatrix is considered to define the DHNM in a way similar to the dynamic read and write noise margins.

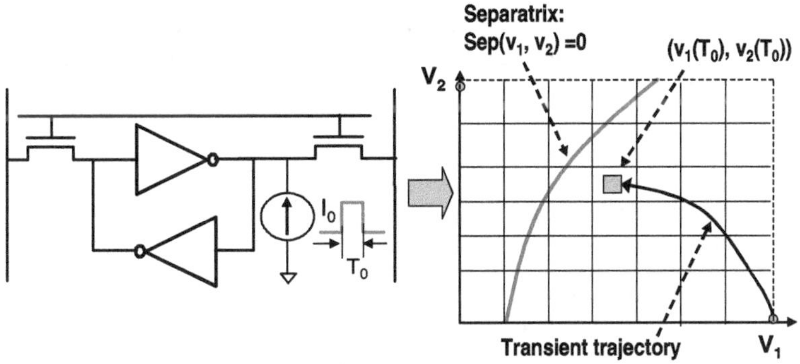

Fig. 6 Definition and characterization of dynamic hold noise margin (DHNM) (from [10] ©[2008] IEEE)

4.4 Relations to Conventional Static Noise Margins

Broadly, the conventional static noise margins [1] are special cases of the dynamic noise margins presented here. For read accesses, it is easy to see that when $T_R = \infty$, the DRNM predicts the stability of the cell in a static sense, i.e. whether or not the state will be flipped if the duration of the read process is infinite. This coincides with checking whether the static read noise margin is greater than zero or not. However, in typical situations the DRNM more realistically captures the disturbance of the cell state resulted from a read access since the access time is finite. In this sense, the static read noise margin is *pessimistic* as it assumes that the read disturbance is not temporary but permanent.

Similarly, in terms of writes, with T_W set to ∞, the DWNM predicts the static write-ability, i.e. whether or not the state can be successfully overwritten if the write process is infinite in duration. Again, this coincides with checking whether the static write noise margin is greater than zero or not. From this angle, for realistic cases, without accounting for the finite duration of the write access, the SNM may provide an *optimistic* estimate for dynamic write-ability. For the state-of-the-art SRAM designs with short access cycles and advanced read/write timing control circuitry, the distinctions between the SNMs and DNMs in read and write reveal the important role of cell nonlinear dynamics in determining dynamic SRAM stability. Finally, a similar conclusion can be drawn upon the distinctions between the DHNM and SNM for hold.

5 Analysis of Dynamic Noise Margins

It becomes clear that the separatrix has an important role in characterizing the stability of static memories and defining dynamic noise margins. We discuss an efficient way of computing the separatrix in the two-dimensional state space and present illustrative examples of dynamic noise margin analysis. In later sections, we discuss inherent numerical issues associated with separatrix computation and provide an extension to higher-dimensional state spaces.

5.1 Computationally Efficient Tracing of Separatrices

To elucidate the computational issue associated with finding the separatrix for a given static memory cell, first consider a brute-force approach as shown in Fig. 7a, which is based upon sampling the state space using a large number of transient simulation runs.

The entire state space of the SRAM cell is sampled using a dense grid to find the boundary between the stability regions of the two stable equilibria. Then, each grid point is used as an initial condition in transistor-level transient simulation. Starting from each initial condition, the SRAM state trajectory may be attracted eventually to one of the stable equilibria. This signifies that the sampled initial condition falls into the stability region of the corresponding equilibrium. If a sufficiently large number of samples are taken, the separation between points falling into the two stability regions forms the separatrix. Computationally, this brute-force state space sampling approach can be quite expensive as the number of transient runs required may be very high. The computational cost is particularly pronounced when process variations are considered in dynamic stability analysis, where a large number of Monte Carlo sampling runs are needed to capture statistical performance variations due to manufacturing fluctuations.

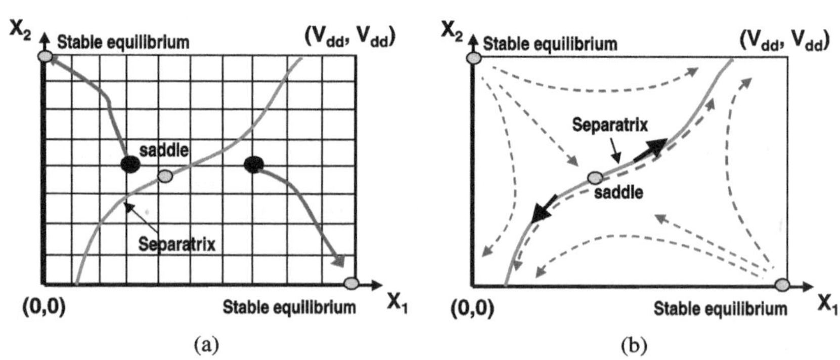

Fig. 7 Separatrix computation: (**a**) brute-force state-space sampling, and (**b**) efficient tracing using system concepts (from [10] ©[2008] IEEE)

To this end, understanding nonlinear system principles not only offers theoretical insights on dynamic stability, but also provides a foundation for a much more computationally efficient alternative. For this, let us denote the stability boundary (separatrix) between the two stability regions as ∂A. For the static memory cells we study in this chapter, the following stability boundary theorem exists [3, 9]:

Theorem 1. *The separatrix (stability boundary) (stability boundary) of the two stable equilibria is the stable manifold of the unstable equilibrium (saddle), that is*

$$\partial A = W^s(x_e^u), \tag{6}$$

where x_e^u *is the unstable equilibrium and it is on the stability boundary* ∂A.

Theorem 1 implies that to find the stability boundary, one shall identify the saddle on the stability boundary and then find its stable manifold. This provides a systematic approach for finding the desired separatrix. Furthermore, from the view point of stable manifold theory, a stable eigenvector of the linearized system at an equilibrium is tangent to its corresponding stable manifold. This implies that by starting in a neighborhood of x_e^u, the desired stable manifold can be found by integrating the system along the direction of the stable eigenvectors backward in time. Integration backward in time prevents the state trajectory to converge to any of the equilibrium points and enforces the trajectory to stay on the stability boundary. The above theoretical results provide a basis for the following highly efficient separatrix tracing algorithm.

Algorithm 1 Fast tracing based separatrix computation algorithm

1: **Input:** given memory cell netlist;
2: **Output:** traced separatrix;
3: Find the saddle (x_e^u) of the cell via DC analysis;
4: Linearize the memory circuit at x_e^u;
5: Find the stable eigenvector, u_s, for the saddle x_e^u: the stable eigenvector corresponds to the stable eigenvalue of the linearized circuit;
6: Choose two initial conditions around x_e^u:
$x^0_{\{1,2\}} = x_e^u \pm \varepsilon u_s$, ε is small;
7: Conduct a transient analysis from each initial condition $x^0_{\{1,2\}}$ for the modified nonlinear dynamical system:

$$f(x(t)) - \frac{d}{dt}q(x(t)) = 0; \tag{7}$$

8: Form the separatrix as:

$$\partial A = \{x \mid x = \phi_M(t, x_1^0) \textbf{ or } \phi_M(t, x_2^0),\ t = [0, T_{\max}]\}, \tag{8}$$

where ϕ_M is the state trajectory of the modified system, and $T_{\max}$ is the maximum duration of the two transient runs.

Note that at *step 5* of Algorithm 1 two transient analysis runs are conducted for the modified dynamic system to trace the separatrix of the original system. To elucidate this point, contrast the phase portrait of this modified system (Fig. 7b) with that of the original one, such as one shown in Fig. 3b. Effectively, integrating the original system in (1), backward in time is reflected by negating the differential term in the modified system in (7). Intuitively, negating the differential term reverses the vector field, as shown in the dashed arrow lines in Fig. 7b. Note that along the separatrix, the direction of the vector field points away from x_e^u. Elsewhere in the state space, the vector field points towards the separatrix. Since the two transient simulations do not start from equilibrium x_e^u, but from a neighborhood of it, the state trajectories will move. Due to the new vector field, they do not converge to the saddle nor the two stable equilibrium points, but are actually forced to stay on the separatrix, making it possible to efficiently find the entire separatrix by using only two transient runs. From a dynamic system point of view, integrating backward in time allows one to trace the desired separatrix efficiently. We discuss the inherent numerical issues in doing so in the next section.

5.2 Illustrative Examples

The presented separatrix tracing and dynamic noise margin analysis techniques have been implemented as part of a SPICE-like simulation environment running on the Linux operating system. We consider a standard 6-T SRAM cell with 1 V supply voltage, as shown in Fig. 8a, under various parameter settings.

5.2.1 Separatrix Tracing

In Fig. 8b, we show the simulated separatrix when the circuit parameters are set up such that the netlist represents an ideal symmetrical SRAM cell. In the figure, the point labeled as "equilibrium point" corresponds to the saddle that sits on the separatrix. As expected, the resulting separatrix is along the 45° line, which verifies

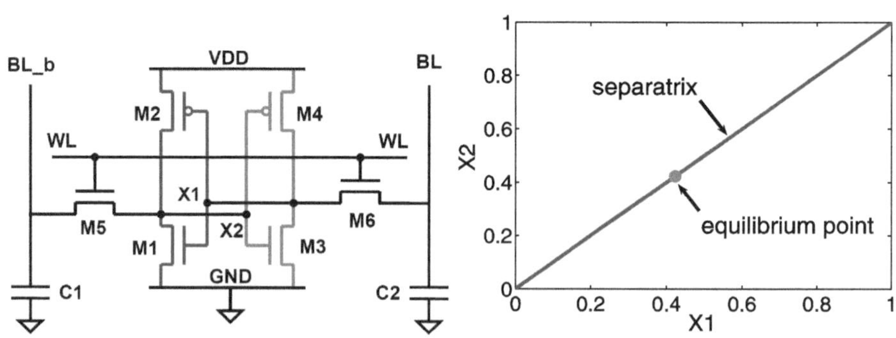

(a) A 6-T SRAM cell. (b) The separatrix of the 6-T cell.

Fig. 8 The separatrix of a symmetric SRAM cell that evenly splits the state space (from [10] ©[2008] IEEE)

the correctness of our fast tracing algorithm. We consider two cells with transistor parameter variations and show the traced separatrice in Fig. 9a and Fig. 9b, respectively. In the former case, the threshold voltages of NMOS transistors M1 and M3 are both increased by 20% while those of PMOS transistors M2 and M4 are decreased by 20%. Since the two cross-coupled inverters are still identical, the separatrix is along the 45° line as before. However, the saddle is shifted to a new location. In the latter case, the effective channel lengths and the threshold voltages of M1 and M2 are both decreased by 30% and those of M3 and M4 are increased by 30%. Due to the mismatch, the separatrix is no longer a straight line and is noticeably distorted.

The average separatrix tracing time is less than 1 minute based on our implementation. To get a sense on the runtime improvement brought by the fast tracing algorithm, we contrast it with the state-space sampling based brute-force method. If a 100×100 grid is used to sample the state space to get down to an accuracy level of 1%, the total sampling based transient analysis runtime is about 38 hours. With this reference, the presented algorithm provides a speedup of more than 2,000X.

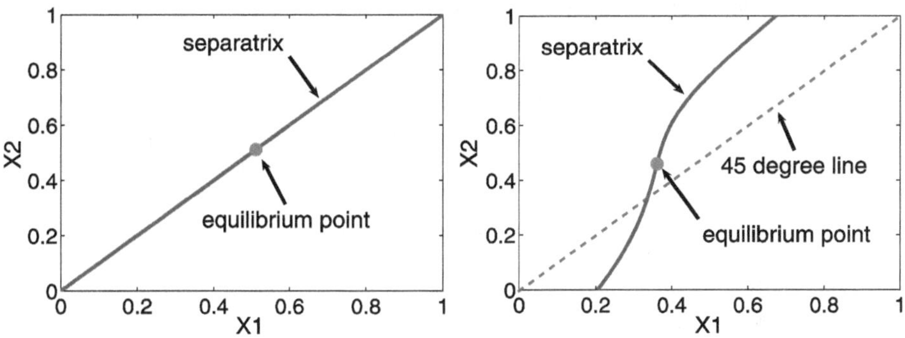

(a) V_T variations only

(b) V_T and L_{eff} variations

Fig. 9 SRAM cell separatrices under device variations (from [10] ©[2008] IEEE)

5.2.2 Dynamic Noise Margin Analysis

With the initial SRAM state set at $x_1 = 1.0$ V and $x_2 = 0$ V, we analyze the dynamic read and write noise margins as defined in the previous sections. We consider an asymmetric 6-T SRAM cell for read DNM analysis. First, the separatrix in hold is traced. A separate transient analysis run is taken to simulate the read access with the two access transistors being turned on. The time at which the state trajectory crosses the separatrix, or T_{across}, is found to be 8.71 ns. Then, consider four wordline turn-on times $T_{R1} = 8.72$ ns, $T_{R2} = 8.70$ ns, $T_{R3} = 8.20$ ns, $T_{R4} = 5.00$ ns. According to the presented dynamic read noise margin definition, the DRNMs for these four cases are found to be −0.01 ns, 0.01 ns, 0.51 ns and 3.71 ns, respectively. Transient simulations are used to verify the DRNM results. The simulation trajectories under the four wordline turn-on times are shown in Fig. 10a. As expected, in the first case a state flip is produced, indicated by a negative read DNM. For other three cases, the DRNM is positive and the SRAM state moves back to the initial value after the read process is completed.

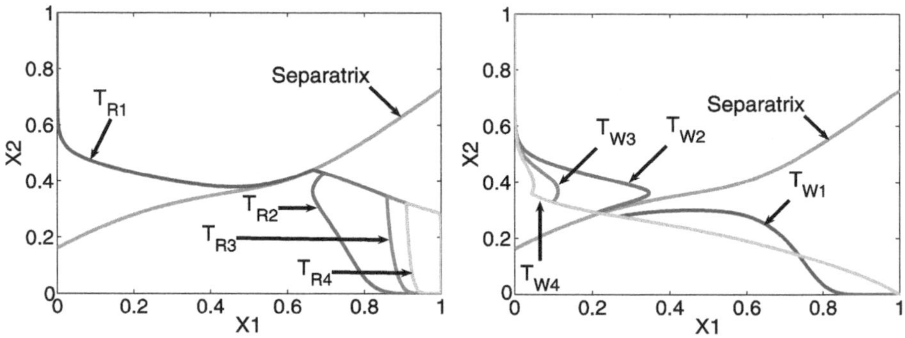

(a) Dynamic read noise margins. (b) Dynamic write noise margins.

Fig. 10 Verification of the simulated DRNMs and DWNMs as functions of access time (from [10] ©[2008] IEEE)

For the same cell, we further analyze dynamic write noise margins with the wordline pulse width (T_{W1} to T_{W4}) set to be 38 ps, 42 ps, 53 ps, 65 ps, respectively. The time it takes for the state trajectory in write to reach the separatrix, T_{across}, is found to be 40 ps. Therefore, the DWNMs are −2 ps, 2 ps, 13 ps and 25 ps, respectively for these four different write access times. Again, we use transient simulations to verify our results, as shown in Fig. 10b. It can be seen that in the first case, no state flip is produced, indicating a write failure, which is correctly predicted by our negative DWNM margin. In all other three cases, the cell is successfully written, consistent to the computed positive DWNMs.

Using the proposed dynamic noise margin analysis technique as a backbone, a computationally efficient parametric DNM analysis approach has been developed [10]. In a multi-dimensional device/design parameter space, this parametric technique can identify the region of parameter combinations, referred to as the acceptance region, in which a specified dynamic noise margin target is met. Correspondingly, other parts of the parameter space is deemed to belong to the failure region. Such capability is instrumental, for instance, in assisting efficient statistical analysis, where the statistical weight of the acceptance region gives the yield of the cell design.

As an example, we consider the nominal threshold voltages for the NMOS and PMOS transistors in the 6-T SRAM cell to be 0.3 V and −0.28 V, respectively. The initial SRAM state is also at $x_1 = 1.0$ V and $x_2 = 0$ V (Fig. 8a). We first consider the V_T variations of transistors M1 and M5 in Fig. 8a, and the acceptance region boundary is set to be:

$$\partial\Omega_{\mathrm{accept}} = \{(V_{\mathrm{th1}}, V_{\mathrm{th5}}) \mid T_{\mathrm{across}}(V_{\mathrm{th1}}, V_{\mathrm{th5}}) = 0.3 \text{ ns}\}, \tag{9}$$

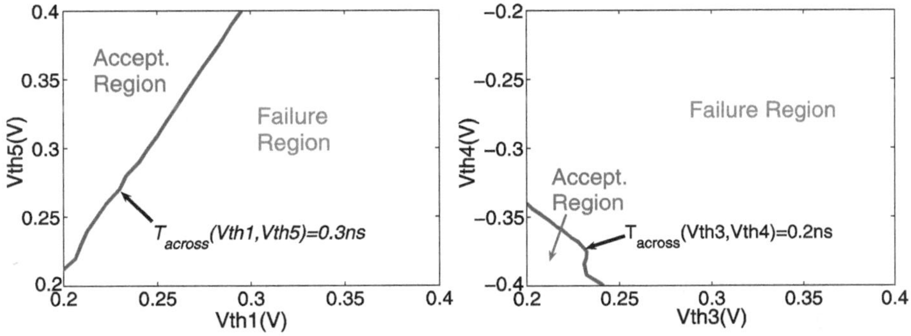

(a) Dynamic read noise margins. (b) Dynamic write noise margins.

Fig. 11 Parametric DRNM and DWNM analysis: boundaries between the acceptance and failure regions in the device parameter space (from [10] ©[2008] IEEE)

where V_{th1} and V_{th5} represent the threshold voltages of transistors M1 and M5. The computed boundary between the acceptance and failure regions is shown in Fig. 11a. Since we have obtained the boundary of acceptance region, the yield of read DNM can be efficiently evaluated. However, if we use Monte Carlo simulation to evaluate the read DNM yield, to cover wide range of variations (e.g. 6σ or beyond), the number of samples required can be huge. One Monte Carlo run on average takes 10 seconds. Assuming one million samples are needed, then a total of 10^7 seconds will be required. For this example, the proposed method takes 63 minutes to complete, providing a runtime speedup of three orders of magnitude.

For parametric DWNM analysis, the threshold voltage variations of M3 and M4 are considered. The acceptance region is set such that T_{across} should be at most 0.2 ns. The computed boundary between the acceptance and failure regions in the parameter space is shown in Fig. 11b. The runtime for this analysis is 71 minutes.

6 Numerical Stability of Separatrix Tracing

In the previous section, integrating backwards in time has been introduced as a way of tracing separatrices of memory cells in an efficient way. However, from a numerical perspective, this process is not necessarily numerically stable. In practice, cares must be taken to ensure the accuracy of the traced separatrices, for instance, by adopting small step sizes. To more fundamentally address this issue, here we suggest to use a recently developed multi-level integration scheme to enhance the stability of backward-in-time numerical integration. The method under discussion is the so-called *reverse projective integration* and its telescopic extension. These methods originated from the numerical analysis community [4, 6] and have been recently exploited for parallel circuit simulation [11].

Simply put, the numerical instability associated with integrating a dynamic system (represented using a DAE) backwards in time stems from the fact that it effectively converts each stable eigenvalue of the system to an unstable one. Hence, one needs to have specific numerical handling to ensure the accumulated errors associated with such eigenvalues are well controlled.

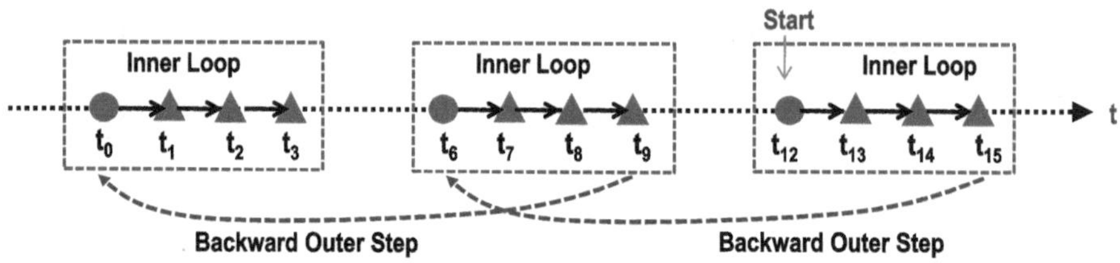

Fig. 12 The reverse projective integration method ($K = 2, M = 9$)

As illustrated in Fig. 12, reverse projective integration employs a combination of 'inner' and 'outer' integration steps. In inner integration steps, an inner integrator integrates $K+1$ steps from t_j to t_{j+K+1} along the forward direction of time using a small step size. The choice of this step size ensures stability and significantly damps the local errors after $K+1$ steps. In the following outer integration step, a larger reverse extrapolation is taken to project backwards from time point t_{j+K+1} to time point $t_{j+K+1-M}$ ($M > K+1$). Note that the reverse projective step only amplifies the accumulated error linearly, which takes place after the sequence of $K+1$ inner integration steps has exponentially damped the error. As a result, the stability of the overall integration scheme is safeguarded even with a large reverse extrapolation step size.

The procedure of the reverse projective integration method RP(K,M) can be summarized as follows: (1) use an inner integrator to integrate for K steps from t_j to t_{j+K} to get solution $x(t_{j+K})$; (2) perform one more inner integration to compute $x(t_{j+K+1})$ from $x(t_{j+K})$; (3) perform a backward extrapolation over M steps using $x(t_{j+K})$ and $x(t_{j+K+1})$ to compute $x(t_{j+K+1-M})$ as

$$x(t_{j+K+1-M}) = Mx(t_{j+K}) - (M-1)x(t_{j+K+1}). \tag{10}$$

In Fig. 12, we show an example of the reverse projective integration with $k = 2$, $M = 9$.

The effective application of reverse projective integration requires known clustering information of system eigenvalues. When the eigenvalues of the system are widely distributed with no clear clustering or the eigenvalue distribution is not known *a priori*, the step size of the outer projective step must be conservatively controlled to ensure stability. This leads to reduced step-size boost via the projective step. To maintain good efficiency for more general cases, the basic concept of reverse projective integration has been generalized to a multi-level telescopic scheme [4]. For numerical integration in the reversed time direction, at each individual projective integration level, a limited step size amplification may be obtained with a relatively small M. However, a significant overall step size amplification may be obtained in an L-level telescopic framework. Such a two-level backward-in-time (reverse) telescopic projective scheme is suggested in Fig. 13.

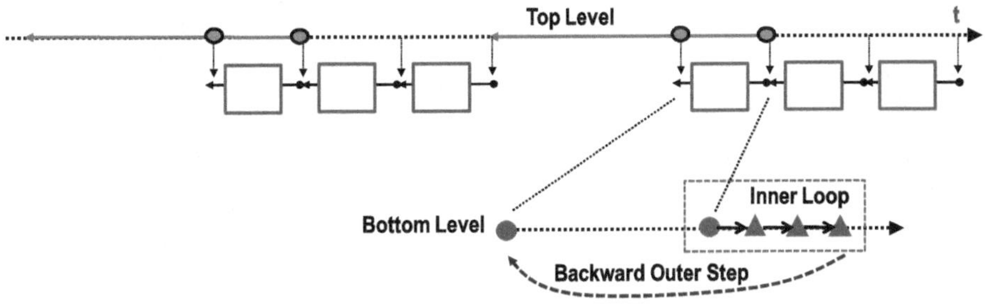

Fig. 13 A two-level reverse telescopic projective integration

7 Extension to Memory Cells Modeled as High-Dimensional Systems

The separatrix tracing algorithm presented earlier can be elegantly and efficiently to SRAM cells that are modeled as a two-dimensional dynamic system, for instance, in the case of pre-layout schematics. However, in addition to the two main internal storage nodes, extracted layout parasitics introduce additional states to the system. This makes it desirable to analyze the separatrix of these higher-dimensional systems in order to fully account for parasitic effects. In principle, it is possible to extend the separatrix tracing algorithm to the N-dimensional state space. For example, it may be achieved by spawning off multiple transient runs around the saddle to trace the $(N-1)$-dimensional separatrix. However, doing so becomes rather cumbersome, as the dimensionality increases.

Recently, a different approach has been demonstrated that deals with separatrix approximation in high-dimensional SRAM memory circuits [12]. In this approach, the tangent of the separatrix, which is an $(N-1)$-dimensional hyperplane is efficiently computed and used for approximating the separatrix to improve the ease and efficiency of dynamic noise margin analysis. According to the stable manifold theory [5], the SRAM separatrix tangent can be computed by linearizing the original nonlinear system at the saddle. The resulting linearized system well approximates the nonlinear dynamics around the saddle, and in general shall have N eigenvalues (eigenvectors) in total with one unstable eigenvalue (eigenvectors). The separatrix is tangent to the $N-1$ stable eigenvectors and hence is spanned by these vectors as illustrated in Fig. 14. While the tangent computation involves eigendecomposition, the computational cost is insignificant as the dimensionality of an extracted SRAM cell is low even with full parasitics.

The computed tangent can be used in two different ways to facilitate dynamic noise margin analysis [12]. Not only is the separatrix tangent close to the separa-

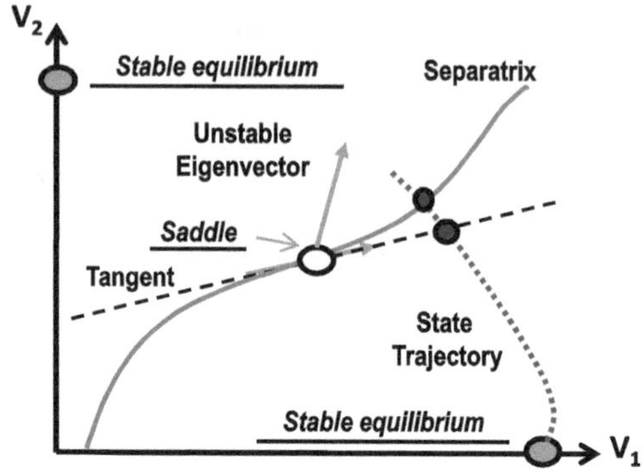

Fig. 14 The separatrix tangent computation for multiple-dimensional memory cells. The tangent-crossing time can be used as an approximate to the separatrix-crossing time, or as a starting point for finding the exact separatrix-crossing time. For ease of illustration, a two-dimensional system is shown

trix around the saddle, it is also found that for practical SRAM designs little error incurs when the separatrix is used in place of the exact separatrix when computing the separatrix-crossing time. This leads to a highly efficient approximate dynamic stability analysis approach. Moreover, an exact algorithm can be developed by applying fast search based iterative refinement. In this case, first, the intersection between the state trajectory and the tangent is obtained as in the approximate algorithm. Around the neighborhood of this intersection, a few transient simulation runs with binarily sized search step sizes are perform to find the exact separatrix-crossing time. As part of a SPICE simulation environment, it has been experimentally demonstrated that such exact dynamic noise margin analysis algorithm is rather efficient. Furthermore,discarding the iterative refinement step as in the approximate algorithm can still produce accurate DNM estimates while providing additional runtime benefits [12].

8 Conclusions

The development of new dynamic stability metrics is stimulated by the fact that the state-of-the-art static memories operate in an increasingly dynamic fashion. This makes the use of traditional static noise margins inappropriate or inaccurate in many cases. Characterizing the dynamic stability of SRAMs and other static memories requires an in-depth understanding of nonlinear cell dynamics. The presented dynamic noise margins are built upon the concept of stability boundary (separatrix). Computationally efficient, numerically robust numerical methods have been discussed for computing or approximating separatrices, which provides a basis for dynamic stability analysis.

Acknowledgements This material is based upon work supported by the US National Science Foundation under Grant No. 0917204 and Grant No. 0747423, and the Semiconductor Research Corporation under Contract 2008-HC-1836.

References

1. J. Lohstroh, E. Seevinck, and J. D. Groot. Worst-case static noise margin criteria for logic circuits and their mathematical equivalence. In *J. of Solid-State Circuits*, (IEEE, Dec 1983), pp. 803–806.
2. E. Seevinck, F. J. List, and J. Lohstroh. Static-noise margin analysis of MOS SRAM cells. In *J. of Solid-State Circuits*, (IEEE, Oct 1987), pp. 748–754.
3. J. Zaborszky, G. M. Huang, B. Zheng, and T. C. Leung. Static-noise margin analysis of MOS SRAM cells. In *Trans. Automatic Control*, (IEEE, Jan 1988), pp. 4–15.
4. C. W. Gear and I. G. Kevrekidis. Telescopic projective methods for parabolic differential equations. In *Physics Letter A*, (vol. 187, 2003), pp. 95–109.
5. S. Wiggins. in *Introduction to Applied Nonlinear Dynamical Systems and Chaos*, (Springer, 2003).
6. C. W. Gear and I. G. Kevrekidis. Computing in the past with forward integration. In *Physics Letter A*, (vol. 321, 2004), pp. 335–343.
7. M. Khellah, Y. Ye, N. S. Kim, D. Somasekhar, G. Pandya, A. Farhang, K. Zhang, C. Webb and V. De. Wordline & bitline pulsing schemes for improving SRAM cell stability in low-Vcc 65 nm CMOS designs. In *Digest of Technical Papers, Symp. on VLSI Circuits*, (IEEE, June 2006), pp. 9–10.
8. H. Pilo et al. An SRAM design in 65 nm and 45 nm technology nodes featuring read and write-assist circuits to expand operating voltage. In *Digest of Technical Papers, Symp. on VLSI Circuits*, (IEEE, June 2006), pp. 15–16.
9. G. M. Huang, W. Dong, Y. Ho, and P. Li. Tracing SRAM separatrix for dynamic noise margin analysis under device mismatch. In *Int. Behavioral Modeling and Simulation Conf.*, (IEEE, 2007), pp. 6–10.
10. W. Dong, P. Li and G. M. Huang. SRAM dynamic stability: theory, variability and analysis. In *Proc. of Int. Conf. on Computer-Aided Design*, (IEEE/ACM, 2008), pp. 378–385.
11. W. Dong and P. Li. Final-value ODEs: stable numerical integration and its application to parallel circuit analysis. In *Proc. of Int. Conf. on Computer-Aided Design*, (IEEE/ACM, 2009), pp. 403–409.
12. Y. Zhang, P. Li and Garng M. Huang. Separatrices in high-dimensional state space: system-theoretical tangent computation and application to SRAM dynamic stability analysis. In *Proc. of Design Automation Conf.*, (IEEE/ACM, 2010), pp. 567–572.

Recycling Circuit Simulation Techniques for Mass-Action Biochemical Kinetics

Jared E. Toettcher, Joshua F. Apgar, Anya R. Castillo, Bruce Tidor and Jacob White

Abstract Many numerical techniques developed for analyzing circuits can be "recycled"—that is, they can be used to analyze mass-action kinetics (MAK) models of biological processes. But the recycling must be judicious, as the differences in behavior between typical circuits and typical MAK models can impact a numerical technique's accuracy and efficiency. In this chapter, we compare circuits and MAK models from this numerical perspective, using illustrative examples, theoretical comparisons of properties such as conservation and invariance of the non-negative orthant, as well as computational results from biological system models.

Jared E. Toettcher
Department of Cellular and Molecular Pharmacology, University of California, San Francisco, California, 94158, USA, e-mail: `jared.toettcher@ucsf.edu`

Joshua F. Apgar
Systems Biology Group, Boehringer Ingelheim Pharmaceuticals, Ridgefield, CT, 06877, USA, e-mail: `apgar@mit.edu`

Anya R. Castillo
Engineering Systems Division, Massachusetts Institute of Technology, Cambridge, MA, 02139, USA, e-mail: `anya.castillo@gmail.com`

Bruce Tidor
Department of Electrical Engineering and Computer Science and the Department of Biological Engineering, Massachusetts Institute of Technology, Cambridge, MA, 02139, USA, e-mail: `tidor@mit.edu`

Jacob White
Department of Electrical Engineering and Computer Science, Massachusetts Institute of Technology, Cambridge, MA, 02139, USA, e-mail: `white@mit.edu`

P. Li et al. (eds.), *Simulation and Verification of Electronic and Biological Systems*,
DOI 10.1007/978-94-007-0149-6_6,

1 Introduction

One important goal of the field of Systems Biology is to develop models of complex biological systems that capture enough of the important mechanisms to be quantitatively accurate and/or to make useful predictions. Such models can serve as a basis for understanding, as an aid in the design of experiments, and as an accelerator for the development of clinical interventions [1, 2]. In this paper, we will focus on the mass-action kinetics (MAK) approach to generating such models, because of its wide use and its strong analogies to circuits [3, 4].

MAK modeling is used to generate a system of differential equations that approximately describes the behavior of a biochemical system. In much the same way, circuit modeling is used to generate a system of differential equations that approximately describes the behavior of an electronic system. In the case of circuits, nearly a half century of intense development has produced an extensive and heavily-optimized library of specialized numerical algorithms and software [5]. Heretofore, far less attention has been devoted to developing numerical techniques for MAK, resulting in opportunities to "recycle" techniques developed for circuits [6].

In this chapter, we examine some of the similarities and differences between circuit and MAK related differential equation systems, primarily from a numerical perspective. Our intent is to accelerate the recycling process, by showing how differences between circuits and MAK models can impact the choice of numerical techniques. The result of ignoring these differences is usually just a loss of efficiency, but in some cases, like the conservation constraints case described below, ignoring the differences can result in numerical techniques that produce erroneous results.

In the next section we compare a three-node circuit example with a three-species MAK example, and use the example to introduce notation. In Sect. 3, we make more general comparisons between circuit and MAK related differential equation systems, and focus on the two issues of conservation constraints and the invariance of the non-negative orthant. In Sect. 4, biochemical system examples are used to demonstrate the effectiveness of exploiting the Kronecker product form for MAK when using scripting languages, and to demonstrate that ignoring conservation constraints can lead to errors when computing oscillator parametric sensitivity.

2 Illustrative Examples

Consider the example circuit and MAK representations in Fig. 1. On the left is a diagram of an example two-resistor, three grounded-capacitor circuit, where the variables of interest are the three circuit node voltages with respect to ground, v_1, v_2 and v_3, and the two electrical currents, i_1 and i_2, through the two resistor elements. On the right of Fig. 1 are two ways of representing a simple MAK example, in which two molecules combine to synthesize a third. On the top-left is a reaction diagram most analogous to a circuit diagram, and on the bottom-left is the equivalent chemical reaction equation. In this MAK example, the variables of interest are the three species concentrations, x_1, x_2 and x_3, and the two mass fluxes, r_1 and r_2, generated by the synthesis and dissociation reactions.

2.1 Circuit Equations

For the circuit example in Fig. 1, the differential equations that describe the trajectories in time of the circuit node voltages can be derived by using *charge-balance* [7]. That is, the time derivative of each node charge is set equal to the signed sum of element currents incident at that node. In this example, the node charges are given by c_1v_1, c_2v_2, and c_3v_3, and satisfy the differential equation system,

$$\mathbf{C}\frac{d}{dt}\mathbf{v}(t) = \mathbf{Ai}(t) \tag{1}$$

where $\mathbf{v}(t) \in \mathbb{R}^3$ is the vector of time-dependent node voltages, $\mathbf{i}(t) \in \mathbb{R}^2$ is the vector of time-dependent element currents, $A \in [-1,0,1]^{3\times 2}$ is an incidence matrix (described below) [7], and $\mathbf{C} \in \mathbb{R}^{3\times 3}$ is a diagonal capacitance matrix. That is,

$$\mathbf{v} = \begin{bmatrix} v_1 \\ v_2 \\ v_3 \end{bmatrix}, \quad \mathbf{i} = \begin{bmatrix} i_1 \\ i_2 \end{bmatrix}, \quad \mathbf{C} = \begin{bmatrix} c_1 & 0 & 0 \\ 0 & c_2 & 0 \\ 0 & 0 & c_3 \end{bmatrix}, \quad \mathbf{A} = \begin{bmatrix} -1 & 0 \\ 1 & 1 \\ 0 & -1 \end{bmatrix}. \tag{2}$$

For the limited set of circuit elements considered herein, element *constitutive equations* relate node voltages, or differences between node voltages, to element currents. In the example in Fig. 1, the element current i_1, through resistor R_1, and element current i_2, through R_2, are given by

$$\mathbf{i} = \mathbf{g}(\mathbf{v}) = \begin{bmatrix} g_1(v_1 - v_2) \\ g_2(v_3 - v_2) \end{bmatrix}, \tag{3}$$

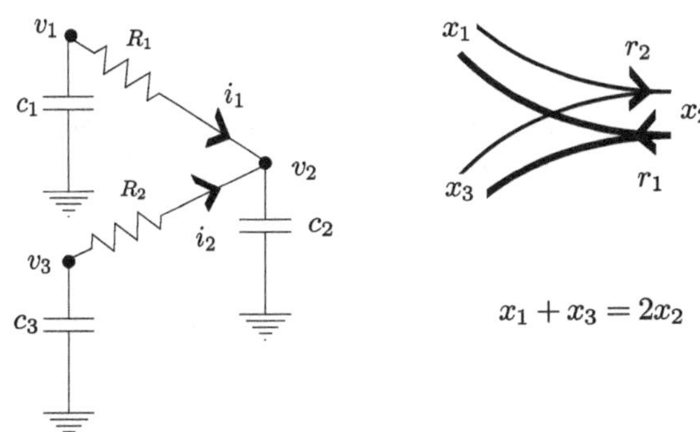

Fig. 1 On the left is the schematic for a two-resistor three grounded capacitor circuit, on the right is a schematic and a chemical reaction equation for a reversible binary reaction

where the function $\mathbf{g}$ maps a vector of node voltages to a vector of element currents. In the linear case, $\mathbf{g}$ becomes a matrix of conductances, as in

$$\mathbf{i} = \mathbf{g}\mathbf{v} = \begin{bmatrix} g_{1,1} & g_{1,2} & g_{1,3} \\ g_{2,1} & g_{2,2} & g_{2,3} \end{bmatrix} \mathbf{v} = \begin{bmatrix} \frac{1}{R_1} & -\frac{1}{R_1} & 0 \\ 0 & -\frac{1}{R_2} & \frac{1}{R_2} \end{bmatrix} \mathbf{v}. \tag{4}$$

If the node voltages are given in volts, and the element currents in amps, then the units for R_1 and R_2 are ohms, denoted Ω. There is humor to be found in the subject of circuits, as the units for conductances are mhos, denoted ℧.

As a last step, we augment the circuit in Fig. 1 with inputs, where the inputs are in the form of source currents injected into each circuit node. Combining these source currents with (1) and (3) yields

$$\mathbf{C}\frac{d}{dt}\mathbf{v}(t) = \mathbf{A}\mathbf{g}\left(\mathbf{v}(t)\right) + \mathbf{i}_s(t) \equiv \mathbf{G}\left(\mathbf{v}(t)\right) + \mathbf{i}_s(t) \tag{5}$$

where $\mathbf{G}$ maps a vector of node voltages to a vector of node currents, and $\mathbf{i}_s(t) \in \mathbb{R}^3$ is the vector of source currents injected at each of the nodes.

Note that (5) are the *nodal* equations [7] for our example circuit. In the linear case, the map $\mathbf{G}$ becomes a matrix, and (5) becomes

$$\mathbf{C}\frac{d}{dt}\mathbf{v}(t) = \begin{bmatrix} -\frac{1}{R_1} & \frac{1}{R_1} & 0 \\ \frac{1}{R_1} & -\frac{1}{R_1} - \frac{1}{R_2} & \frac{1}{R_2} \\ 0 & \frac{1}{R_2} & -\frac{1}{R_2} \end{bmatrix} \mathbf{v}(t) + \mathbf{i}_s(t) = \mathbf{G}\mathbf{v}(t) + \mathbf{i}_s(t). \tag{6}$$

2.2 MAK Equations

For the MAK example in Fig. 1, the differential equations that describe the trajectories in time of the species concentrations, x_1, x_2 and x_3, are derived using *mass balance* [8]. That is, the species concentrations satisfy the differential equation system,

$$\frac{d}{dt}\mathbf{x}(t) = \mathbf{S}\mathbf{r}(t) \tag{7}$$

where $\mathbf{x}(t) \in \mathbb{R}^3$ is the vector of time-dependent species concentrations, $\mathbf{r}(t) \in \mathbb{R}^2$ is the vector of time-dependent reaction fluxes, and $\mathbf{S} \in \mathbb{Z}^{3\times 2}$ is referred to as the stoichiometric matrix (described below) [3]. That is,

$$\mathbf{x} = \begin{bmatrix} x_1 \\ x_2 \\ x_3 \end{bmatrix}, \quad \mathbf{r} = \begin{bmatrix} r_1 \\ r_2 \end{bmatrix}, \quad \mathbf{S} = \begin{bmatrix} -1 & 1 \\ 2 & -2 \\ -1 & 1 \end{bmatrix}. \tag{8}$$

For the first-, and second-order reactions considered herein, the associated *constitutive equations* relate a species concentration, or the product of two species concentrations, to a reaction flux. In this example, the rate of synthesis (aka the forward reaction), where a molecule of species 1 combines with a molecule of species 3 to form two molecules of species 2, is denoted by flux r_1, and the rate of the dissociation (aka the reverse reaction), where species 2 breaks down into species 1 and 3, is denoted by flux r_2. The fluxes are related to the species concentrations by

$$\mathbf{r} = \begin{bmatrix} k_f x_1 x_3 \\ k_r x_2 \end{bmatrix} = \mathbf{k}^{(0)}\mathbf{x} + \mathbf{k}^{(1)}\mathbf{x}\otimes\mathbf{x} \tag{9}$$

where k_f and k_r are reaction rate constants, $\mathbf{k}^{(0)} \in \mathbb{R}_+^{2\times 3}$ is the matrix of rate constants for first-order reactions and $\mathbf{k}^{(0)}_{2,3} = k_r$ is its only nonzero entry, $\mathbf{k}^{(1)} \in \mathbb{R}_+^{2\times 9}$ is the matrix of rate constants for second-order reactions and $\mathbf{k}^{(1)}_{1,3} = k_f$ is its only nonzero entry, and $\mathbf{x}\otimes\mathbf{x}$ denotes the Kronecker product of $\mathbf{x}$ with itself [9]. In this case $\mathbf{x}\otimes\mathbf{x}$ is given by (writing in transpose form to save space),

$$\mathbf{x}\otimes\mathbf{x} = \begin{bmatrix} x_1x_1 & x_1x_2 & x_1x_3 & x_2x_1 & x_2x_2 & x_2x_3 & x_3x_1 & x_3x_2 & x_3x_3 \end{bmatrix}^T. \tag{10}$$

Three aspects of (9) are worth note. First, the reaction rate constants k_f and k_r do not share the same units. If the species concentrations are specified in moles per liter (one mole $= 6.022\times 10^{23}$ molecules), then k_r is in $\frac{1}{seconds}$, and k_f is in $\frac{liter}{mole-seconds}$. Second, in what must seem a marvel of inefficiency, we represented the single second-order reaction term, $k_f x_1 x_3$, with the product of a 3×9 matrix $\mathbf{k}^{(1)}$ with the 9-entry long result of $\mathbf{x}\otimes\mathbf{x}$. Nevertheless, the Kronecker product form has both notational and computational value, as we show in Sect. 4. Finally, from (10), both the third and the seventh components of $\mathbf{x}\otimes\mathbf{x}$ are identical, $x_1x_3 = x_3x_1$. This duplication of the x_1x_3 term means that there could be two nonzero entries for $\mathbf{k}^{(1)}$ in (9), with the constraint that $\mathbf{k}^{(1)}_{1,3} + \mathbf{k}^{(1)}_{1,7} = k_f$. In practice, a good strategy is to give equivalent terms equal weight, to help preserve symmetry. So, $\mathbf{k}^{(1)}_{1,3} = \mathbf{k}^{(1)}_{1,7} = \frac{k_f}{2}$.

Paralleling the circuit example, the last step is to augment the MAK system with inputs, where inputs in the MAK case take the form of production rates for each of the species. It should be noted that these input production rates can arise from modeling two very different phenomena, and both may be present in a single model. If the MAK model represents the behavior of a biological signaling process in the cell, then the production rate for a given species may be a concise way to model the cell's complex process for manufacturing that species. Alternatively, the input production rate for a species may be the effect of a researcher adding chemical species into the cell's environment. We call attention to this distinction because the label "input" suggests that one has complete control, when it is more likely the case that the researcher and the cell are using production rates to compete for control of the cell's behavior (as in chemotherapy, for example).

Combining these production rates with (7) and (9) yields

$$\frac{d}{dt}\mathbf{x}(t) = \mathbf{S}\big(\mathbf{k}^{(0)}\mathbf{x} + \mathbf{k}^{(1)}\mathbf{x}\otimes\mathbf{x}\big) + \mathbf{u}(t) \equiv \mathbf{K}^{(0)}\mathbf{x} + \mathbf{K}^{(1)}\mathbf{x}\otimes\mathbf{x} + \mathbf{u}(t), \tag{11}$$

where $\mathbf{K}^{(0)} \in \mathbb{R}^{3\times 3}$, $\mathbf{K}^{(1)} \in \mathbb{R}^{3\times 9}$, and $\mathbf{u}(t) \in \mathbb{R}^3$ is the vector of species production rates. For the specific example in Fig. 1,

$$\frac{d}{dt}\mathbf{x}(t) = \begin{bmatrix} k_r x_2 - k_f x_1 x_3 \\ -2k_r x_2 + 2k_f x_1 x_3 \\ k_r x_2 - k_f x_1 x_3 \end{bmatrix}. \tag{12}$$

Note that the form of the MAK equation system in (11) is strikingly similar to the form of the circuit equation system in (5).

2.3 Leakage and Degradation

In a circuit model of an electronic system, a capacitor may be a model for the electric field interaction between a pair of conductors separated by an insulator. If the insulator is imperfect, the physical capacitor may *leak*, in which case the appropriate circuit model is a capacitor in parallel with a leakage resistor. There is an analogy to leakage resistors in MAK models of biochemical processes in cells. In a physical cell, molecules float through the cytoplasm and bind to form larger complexes, or disassociate into constituent molecules, or are transformed by enzymatic interactions. In a MAK model, binding, disassociation, and enzymatic transformation are all represented as reactions that change concentrations of molecular species. If, in addition, the cytoplasm contains enzymes that completely *degrade* certain molecules, that process is easily included in the MAK model, by adding degradation reactions.

As an example, suppose capacitor c_2 in Fig. 1 leaks. Then the current through the associated leakage resistor becomes a third current in the circuit, lengthening the vector of time-dependent element currents by one. One column must be added to the incidence matrix in (2),

$$\mathbf{A} = \begin{bmatrix} -1 & 0 & 0 \\ 1 & 1 & -1 \\ 0 & -1 & 0 \end{bmatrix} \tag{13}$$

and a constitutive equation for the leakage resistor must be added to (3),

$$\mathbf{i} = \mathbf{g}(\mathbf{v}) = \begin{bmatrix} g_1(v_1 - v_2) \\ g_2(v_3 - v_2) \\ \frac{1}{R_L} v_2 \end{bmatrix}, \tag{14}$$

where now $\mathbf{i}(t) \in \mathbb{R}^3$.

As an analogous MAK example, suppose species 2 in the MAK model in Fig. 1 degrades. The degradation reaction will be a third reaction in MAK model, lengthening the vector of time-dependent fluxes by one. One column must be added to the stoichiometric matrix in (8),

$$\mathbf{S} = \begin{bmatrix} -1 & 1 & 0 \\ 2 & -2 & -1 \\ -1 & 1 & 0 \end{bmatrix}. \tag{15}$$

and a constitutive equation must be added to (9) for the degradation reaction

$$\mathbf{r} = \begin{bmatrix} k_f x_1 x_3 \\ k_r x_2 \\ k_d x_2 \end{bmatrix} = \mathbf{k}^{(0)}\mathbf{x} + \mathbf{k}^{(1)}\mathbf{x} \otimes \mathbf{x} \tag{16}$$

where k_d is the degradation rate constant, $\mathbf{k}^{(0)} \in \mathbb{R}_+^{3\times 3}$ and now has two non-zero terms, $\mathbf{k}^{(0)}_{2,2} = k_r$ and $\mathbf{k}^{(0)}_{3,2} = k_d$. Finally, $\mathbf{k}^{(1)} \in \mathbb{R}_+^{3\times 9}$ but is still zero except for $\mathbf{k}^{(1)}_{1,3} = k_f$.

3 System Comparisons

In Sect. 2, we used simple examples to demonstrate that circuits and MAK models generate differential equation systems that share some common structure, though we provided scant insight into the numerical issues. For instance, node-voltage time derivatives are proportional to signed sums of element currents, a seemingly perfect parallel to our MAK examples, where species-concentration time derivatives are proportional to signed sums of reaction fluxes. Yet, there is a key difference that must be reflected in the choice of numerical techniques. For the circuit examples, (5) is valid regardless of the sign of the node voltages, but for the MAK examples, (11) is *only* valid if the species concentrations are nonnegative.

In this section, we compare circuit and MAK differential equation systems in a little more generality, but restrict consideration to representative classes of circuits and MAK models, where these classes are chosen to best illuminate the numerical differences.

3.1 Circuit Case

Almost any system of differential equations can be matched precisely to an equivalent circuit, and such transformations have been shown to provide insight into a variety of physical and implementation issues [4, 10]. The focus of this paper is on comparing numerical issues for typical MAK models and circuits, so we consider a restricted class that is representative of typical circuits, yet can be insightfully compared to MAK models. The particular restrictions are:

- the only storage elements in the circuit are positive linear capacitors with one terminal connected to ground (grounded-capacitor).
- all non-capacitor element currents can be written as explicit functions of node voltages (voltage-control).

- all non-capacitor element currents flow between pairs of nodes, or from a node to ground, and the current always flows *from* the higher voltage node *to* the lower voltage node, or is zero if the two node voltages are equal (passivity).
- the only inputs to the circuit are currents injected into circuit nodes (current-source-input).

The above restrictions are not as limiting as they might seem. Voltage sources can be modeled using Norton equivalents [7]; most transistor models satisfy the voltage-control and passivity constraints [11, 12]; and the grounded-capacitor constraint is a common simplification used to accelerate the simulation of large digital integrated circuits. On the other hand, to accurately simulate analog integrated circuits, it would be necessary to allow both floating and nonlinear capacitors, though such a generalization would obfuscate the comparison to MAK.

If an n-node-n_e-element circuit satisfies the above constraints, then the node voltages will satisfy a differential equation of the form

$$\mathbf{C}\frac{d}{dt}\mathbf{v}(t) = \mathbf{A}\mathbf{g}(\mathbf{v}(t)) + \mathbf{i}_s(t) \equiv \mathbf{G}(\mathbf{v}(t)) + \mathbf{i}_s(t), \tag{17}$$

where $\mathbf{v}(t) \in \mathbb{R}^n$ is the vector of node voltages at time t, $\mathbf{C} \in \mathbb{R}^{n\times n}$ is a diagonal capacitance matrix (diagonal because all capacitors are grounded), $\mathbf{g}$ maps n node voltages to n_e element currents, $\mathbf{G}$ maps n node voltages to n node currents, and $\mathbf{i}_s(t) \in \mathbb{R}^n$ is the vector of injected node currents at time t. The incidence matrix, $\mathbf{A} \in [-1,0,+1]^{n\times n_e}$, has one row for each circuit node and one column for each element current. If the mth element current flows from node i to node j when $\mathbf{g}_m(\mathbf{v}(t)) > 0$, then the only two non-zero values in the mth column of $\mathbf{A}$ are $\mathbf{A}_{i,m} = -1$ and $\mathbf{A}_{j,m} = 1$. If the mth element current flows between node i and ground, then there is only one nonzero entry in the mth column of $\mathbf{A}$, $\mathbf{A}_{i,m}$. If $\mathbf{g}_m(\mathbf{v}(t)) > 0$ when current flows toward ground, then $\mathbf{A}_{i,m} = -1$, and $\mathbf{A}_{i,m} = 1$ otherwise.

3.2 MAK Case

Differential equation models for cell signaling and transduction often contain equations that are not strictly MAK, but include quasi-steady-state approximations. For example, enzyme kinetics are often approximated using the Michaelis-Menten or the Hill equations, though we will restrict our focus to strictly MAK models [13].

For any physically reasonable MAK model,

- all the input production rates and all the reaction rate constants are non-negative,
- the flux of any reaction that consumes species i is exactly zero whenever the concentration of species i is exactly zero.

We make note of these conditions, even though a model that violates either condition would make no physical sense, as their specific forms arise in theorems of subsequent subsections.

If an n-species MAK model has only first- and second-order reactions, and if the only inputs to the model are production rates, then the species concentrations satisfy a differential equation of the form

$$\frac{d}{dt}\mathbf{x}(t) = \mathbf{S}\big(\mathbf{k}^{(0)}\mathbf{x} + \mathbf{k}^{(1)}\mathbf{x}\otimes\mathbf{x}\big) + \mathbf{u}(t) \equiv \mathbf{K}^{(0)}\mathbf{x} + \mathbf{K}^{(1)}\mathbf{x}\otimes\mathbf{x} + \mathbf{u}(t), \tag{18}$$

where n_r is the number of reactions, $\mathbf{x}(t) \in \mathbb{R}^n$ is the vector of species concentrations at time t, $\mathbf{k}^{(0)} \in \mathbb{R}_+^{n_r \times n}$ is the matrix of rate constants for first-order reactions, $\mathbf{k}^{(1)} \in \mathbb{R}_+^{n_r \times n^2}$ is the matrix of rate constants for second-order reactions, and $\mathbf{u}(t) \in \mathbb{R}^n$ is the vector of species production rates. The stoichiometric matrix, $\mathbf{S} \in \mathbb{Z}^{n \times n_r}$, has one row for each species and one column for each reaction. If the mth reaction is a degradation reaction involving species i, then $\mathbf{S}_{i,m} = -1$ is the only non-zero entry in column m. If the mth reaction is first-order, and consumes α molecules of species i to produce γ molecules of species j then $\mathbf{S}_{i,m} = -\alpha$ and $\mathbf{S}_{j,m} = \gamma$. If the mth reaction is a second-order synthesis reaction, and consumes α molecules of species i and β molecules of species j to produces γ molecules of species k, then $\mathbf{S}_{i,m} = -\alpha$, $\mathbf{S}_{j,m} = -\beta$, and $\mathbf{S}_{k,m} = \gamma$. Finally, if the mth reaction is a disassociation reaction, and consumes γ molecules of species k to produce α molecules of species i and β molecules of species j, then $\mathbf{S}_{i,m} = \alpha$, $\mathbf{S}_{j,m} = \beta$, and $\mathbf{S}_{k,m} = -\gamma$.

3.3 Conservation Constraints

Let us consider a general setting first. Suppose $\mathbf{x}(t) \in \mathbb{R}^n$, $t \geq 0$, satisfies the nonlinear differential equation system

$$\frac{d}{dt}\mathbf{x}(t) = \mathbf{f}(\mathbf{x}(t)) + \mathbf{u}(t), \tag{19}$$

where $\mathbf{u}(t) \in \mathbb{R}^n$ and $\mathbf{f} : \mathbb{R}^n \to \mathbb{R}^n$. We say (19) has n_c *conservation constraints* if there exists a single rank $n_c < n$ matrix, $\mathbf{W} \in \mathbb{R}^{n \times n_c}$, such that any $\mathbf{x}(t) \in \mathbb{R}^n$, $t \geq 0$ that satisfies (19) also satisfies

$$\mathbf{W}^T\mathbf{x}(t) = \mathbf{W}^T\mathbf{x}(0). \tag{20}$$

Given our initial-condition independent definition of conservation constraints, it follows that for any $\mathbf{x}(0)$,

$$\mathbf{W}^T\frac{d}{dt}\mathbf{x}(0) = \mathbf{W}^T\left(\mathbf{f}(\mathbf{x}(0)) + \mathbf{u}(0)\right) = 0. \tag{21}$$

Conservation conditions can usually be determined by analyzing the incidence and stoichiometric matrices, though not always. Also, conservation conditions are common in MAK but quite uncommon in practical circuits. In this subsection we will examine these issues as well as the impact of conservation conditions on numerical methods for time integration (almost none) and computing steady-state (significant). The most critical numerical ramification of conservation conditions, their impact on correct computation of parametric sensitivities, will be the subject of the next section.

3.3.1 Incidence and Stoichiometric Matrices

For circuits and MAK, there are parallel sufficient conditions for conservation constraints [3, 7].

Theorem 1. *If the transpose of the incidence matrix,* $\mathbf{A}$ *from (17), has an* n_c*-dimensional null space and the input,* $\mathbf{i}_s(t)$*, is orthogonal to that null space for all* $t > 0$*, then (17) has at least* n_c *conservation constraints.*

Theorem 2. *If the transpose of the stoichiometric matrix,* $\mathbf{S}$ *from (18), has an* n_c*-dimensional null space and the input,* $\mathbf{u}(t)$*, is orthogonal to that null space for all* $t > 0$*, then (18) has at least* n_c *conservation constraints.*

In Sect. 3.4, we show by example that a theorem about circuits and a theorem about MAK models can have comparable statements, yet have proofs that depend on remarkably dissimilar mechanisms. In this case, though, the proofs are structurally identical.

Proof. Multiplying (17) and (18) by the transposes of rank n_c matrices $\mathbf{W}_A, \mathbf{W}_S \in \mathbb{R}^{n\times l}$, respectively, yields

$$\mathbf{W}_A^T \frac{d}{dt}\mathbf{v}(t) = \mathbf{W}_A^T \mathbf{A}\mathbf{g}(\mathbf{v}(t)) + \mathbf{W}_A^T \mathbf{i}_s(t), \tag{22}$$

and

$$\mathbf{W}_S^T \frac{d}{dt}\mathbf{x}(t) = \mathbf{W}_S^T \mathbf{S}\left(\mathbf{k}^{(0)}\mathbf{x} + \mathbf{k}^{(1)}\mathbf{x}\otimes\mathbf{x}\right) + \mathbf{W}_S^T \mathbf{u}(t). \tag{23}$$

If the conditions of Theorem 1 are satisfied, then there exists a choice for $\mathbf{W}_A$ such that $\mathbf{W}_A^T\mathbf{A} = 0$ and $\mathbf{W}_A^T\mathbf{i}_s(t) = 0$ for all $t \geq 0$. Similarly, if the conditions of Theorem 2 are satisfied, then there exists a choice for $\mathbf{W}_S$ such that $\mathbf{W}_S^T\mathbf{S} = 0$ and $\mathbf{W}_S^T\mathbf{u}(t) = 0$ for all $t \geq 0$. For these choices of $\mathbf{W}_A$ and $\mathbf{W}_S$, (22) and (23) becomes

$$\frac{d}{dt}\left(\mathbf{W}_A^T\mathbf{v}(t)\right) = 0 \tag{24}$$

and

$$\frac{d}{dt}\left(\mathbf{W}_S^T\mathbf{x}(t)\right) = 0, \tag{25}$$

proving the two theorems. □

There is an easy way to see why Theorems 1 and 2 are only sufficient conditions. In the circuit case, consider adding n circuit elements, one from each node to ground. If the constitutive relations for the current in these added elements are $\mathbf{i}_j = 0.0 * \mathbf{v}_j$, $j \in [1, \ldots, n]$ (zero conductance or infinite resistance), then the behavior of the augmented circuit is unchanged, but the original incidence matrix will be appended by an n-dimensional identity matrix, and the transpose of the appended incidence matrix will no longer have a null space. The parallel example in the MAK case is, add a degradation reaction for each species in an n-species model, but assign each degradation reaction a rate constant of zero. Clearly the behavior of the augmented MAK model is unchanged, but the original stoichiometric matrix will be appended with an n-dimensional identity matrix, and the transpose of the appended stoichiometric matrix will no longer have a null space.

3.3.2 Impact of Conservation Constraints

There is usually a simple physical interpretation of a conservation constraint in a circuit or MAK model. In MAK models, it is common to have subsets of species that interact through reversible reactions, and have neither production rates nor degradation reactions. For such subsets of species, there must be a weighted sum of concentrations that is time-invariant, thereby forming a conservation constraint. For the class of circuits we are considering in this paper, any conservation constraint is likely to be associated with an isolated subset of nodes. By isolated subset, we mean there are no source currents to nodes in the subset, and no element currents from subset nodes to ground or to nodes outside the subset.

For the classes of circuits we consider here, this isolation condition is equivalent to creating a cut-set of capacitors, an *extremely uncommon* case in practical circuits. A particularly simple way to capture why circuits do not typically generate conservation conditions is summarized in the following theorem whose proof can be found in many circuit theory textbooks [7].

Theorem 3. *For a circuit satisfying the restrictions in Sect. 3.1, if, from every node in the circuit, there is a path of element currents to ground (excluding the capacitor currents), then the transpose of the associated incidence matrix has no null space.*

The conservation constraints given in (20) generate different problems than the kind of energy conservation constraints that led to the development of symplectic integration methods. As a quick way to see why the conservation constraints considered here have little impact on the choice of time-integration methods, at least in the non-parametric case, consider applying one step of the forward-Euler algorithm to (19). The equation for $\mathbf{x}$ after one forward-Euler time-step is

$$\mathbf{x}(\delta t) = \mathbf{x}(0) + \delta t\,(\mathbf{f}(\mathbf{x}(0)) + \mathbf{u}(t)) \tag{26}$$

where δt is the time-step, and we are being careless and using $\mathbf{x}(\delta t)$ even though we are computing an approximation.

Suppose the exact solutions to (19) satisfy (20), and we multiply (26) through by the transpose of the $\mathbf{W}$ from (20). The result is

$$\mathbf{W}^T \mathbf{x}(\delta t) = \mathbf{W}^T \mathbf{x}(0) + \delta t \mathbf{W}^T (\mathbf{f}(\mathbf{x}(0)) + \mathbf{u}(0)). \tag{27}$$

Using (21), we can simplify (27),

$$\mathbf{W}^T \mathbf{x}(\delta t) = \mathbf{W}^T \mathbf{x}(0), \tag{28}$$

and we have verified that one step of forward-Euler preserves the conservation constraints regardless of step size.

There are a variety of technical subtleties to extending this first-step analysis to subsequent time-steps for a general system of differential equations. Those subtleties disappear if the conservation constraints are null spaces of A and we are using forward-Euler to solve (17), or the conservation constraints are null spaces of S and we are using forward-Euler to solve (18). For these cases, forward-Euler computed solutions will always satisfy our conservation constraints at every time-step. The same could be said for almost any multi-step method, though if the method is implicit, the implicit equations must be solved accurately to ensure conservation.

Conservation constraints are more problematic when computing steady-states. For (19) with a time-independent input, if $\mathbf{x}^*$ satisfies

$$\mathbf{f}(\mathbf{x}^*) + \mathbf{u} = 0, \tag{29}$$

then $\mathbf{x}^*$ is a steady-state, also known as an equilibrium point.

The usual approach to computing $\mathbf{x}^*$ is to use some variant of the iterative Newton's method [14]. Applied to solving (29), the kth iteration of a basic Newton method is

$$\frac{\partial \mathbf{f}}{\partial x}(\mathbf{x}^k)\left(\mathbf{x}^{k+1} - \mathbf{x}^k\right) = -\mathbf{f}(\mathbf{x}^k) - \mathbf{u}, \tag{30}$$

where $\mathbf{x}^{k+1}$ and $\mathbf{x}^k$ are the $k+1^{\text{th}}$ and k^{th} Newton iterates, and if the iteration converges, $\lim_{k \to \infty} \mathbf{x}^k \to \mathbf{x}^*$.

If there are conservation constraints, then Newton's method will have difficulties because the Jacobian matrix, $\frac{\partial \mathbf{f}}{\partial x}(\mathbf{x}^k)$, will be singular. In particular, if there are n_c conservation constraints, the left null space of the Jacobian matrix will have rank n_c. To show this, we substitute $\mathbf{x}^k$ for $\mathbf{x}(0)$ in (21) and differentiate with respect to $\mathbf{x}^k$,

$$\frac{d}{dx}\left(\mathbf{W}^T \left(\mathbf{f}(\mathbf{x}^k) + \mathbf{u}\right)\right) = \mathbf{W}^T \frac{\partial \mathbf{f}}{\partial x}(\mathbf{x}^k) = \frac{d}{dx} 0 = 0. \tag{31}$$

Since the right-hand side in (30) is orthogonal to $\mathbf{W}$, $\mathbf{W}^T(\mathbf{f}(\mathbf{x}^k) + \mathbf{u}) = 0$, almost any approach to solving singular systems, such as QR [15], will succeed in resolving this problem with Newton's method. For example, one can ensure that $\mathbf{W}^T \mathbf{x}^* = \mathbf{W}^T \mathbf{x}^0$ (the initial guess is "conserved") by replacing (30) with the non-square system

$$\begin{bmatrix} \frac{\partial \mathbf{f}}{\partial x}(\mathbf{x}^k) \\ W^T \end{bmatrix} \left[\left(\mathbf{x}^{k+1} - \mathbf{x}^k\right)\right] = \begin{bmatrix} -\mathbf{f}(\mathbf{x}^k) - \mathbf{u} \\ 0 \end{bmatrix}. \tag{32}$$

3.4 The Non-negative Orthant

The non-negative orthant, denoted $\mathbb{R}^n_+$, is the subset of $\mathbb{R}^n$ of all vectors with only non-negative components. Since species concentrations make physical sense only if they are non-negative, one expects that vectors of species concentrations will always be in the non-negative orthant. But, one might not expect that there is also a non-negative orthant property for our restricted class of circuits. And finally, one might actually be surprised to learn that the MAK non-negative orthant properties are *more fragile* than the circuit non-negative orthant properties.

As a step toward analyzing why species concentrations, and node voltages under certain circumstances, remain in the non-negative orthant, consider the sign properties of reaction fluxes and element currents. For circuits, element currents can have positive or negative values, even if all the node voltages are non-negative. Also, the signs of the entries of the incidence matrix, in (17), have no fixed physical meaning. These signs just indicate how to relate the signs of the element currents to current flow directions. The situation is quite different in MAK modeling. Reaction fluxes are always non-negative, and the signs of the entries of the stoichiometric matrix, in (18), do have a fixed physical meaning. Positive entries indicate that the reaction associated with the stoichiometric matrix column produces the species associated with the stoichiometric matrix row, whereas negative entries indicate that the reaction consumes the species. A more precise statement is given in the following Lemma.

Lemma 1. *Given (18) is generated from a MAK model that satisfies the conditions at the beginning of Sect. 3.2, if* $\mathbf{x}(t) \in \mathbb{R}^n_+$*, then* $\mathbf{r}(t) \in \mathbb{R}^n_+$*, where* $\mathbf{r}(t)$ *is the vector of reaction fluxes.*

Proof. The vector of reaction fluxes, from (18), is given by

$$\mathbf{r}(t) = \mathbf{k}^{(0)}\mathbf{x}(t) + \mathbf{k}^{(1)}\left(\mathbf{x}(t) \otimes \mathbf{x}(t)\right). \tag{33}$$

By assumption, the rate constants and the species concentrations are all non-negative, so each component of $\mathbf{r}(t)$ is a sum of products of non-negative quantities. Since each component of $\mathbf{r}(t)$ is non-negative, $\mathbf{r}(t) \in \mathbb{R}^n_+$. □

A next step in analyzing why species concentrations remain in the non-negative orthant, we consider what happens to species concentrations at the boundary, to see if they "hop the fence".

Theorem 4. *If (18) is generated from a MAK model that satisfies the conditions at the beginning of Sect. 3.2, and if* $\mathbf{x}(t)$ *in (18) is in the non-negative orthant, then for all i such that* $\mathbf{x}_i(t) = 0$, $\frac{d}{dt}\mathbf{x}_i(t) \geq 0$.

Proof. First note that $\frac{d}{dt}\mathbf{x}_i(t)$ is equal to the ith input production rate *plus* the sum of fluxes for reactions that produce species *i minus* the sum of fluxes for reactions that consume species *i*. From the conditions at the beginning of Sect. 3.2, the ith input production rate is non-negative. From Lemma 1 and the assumption that $\mathbf{x}(t) \in$

$\mathbb{R}^n_+$, all reaction fluxes are non-negative, so the sum of fluxes for reactions that produce species i must be non-negative. Combining the conditions at the beginning of Sect. 3.2 with the assumption that $\mathbf{x}_i(t) = 0$, any reaction that consumes species i has zero flux, and sum of these fluxes is also zero. Finally, since $\frac{d}{dt}\mathbf{x}_i(t)$ is equal to the sum of two non-negative quantities minus a quantity that is precisely zero, it must be true that $\frac{d}{dt}\mathbf{x}_i(t) \geq 0$. □

The formal statement of the informal presumption that species concentrations remain in the non-negative orthant is given in the following theorem.

Theorem 5. *If (18) is generated from a MAK model conforming to the conditions at the beginning of Sect. 3.2, and a given trajectory* $\mathbf{x}(t)$, $t \geq 0$ *satisfies (18), and* $\mathbf{x}(0) \in \mathbb{R}^n_+$, *then* $\mathbf{x}(t) \in \mathbb{R}^n_+$ *for all* $t \geq 0$.

Theorem 5 follows from Theorem 4. If the initial condition for all the species concentrations is non-negative, then, by Theorem 4, no single species concentration, or set of species concentrations, can be the first to "hop the fence". Or, one can think more physically. If the species concentrations are all non-negative, and one of the species has a concentration of zero, any reaction that consumes that species must have zero flux. Without a consumption reaction with non-zero flux, that species concentration can not decrease, and can not escape the non-negative orthant.

We will not present a formal proof of Theorem 5, as there are some distracting technical details, and we are unlikely to do justice to the excellent expositions in [16] and [17]. Instead, we demonstrate the numerical fragility of the theorem. Consider the example of Sect. 2, with $k_r = 0$ and $k_f = 1.0$. Then (11) becomes

$$\frac{d}{dt}\begin{bmatrix} x_1 \\ x_2 \\ x_3 \end{bmatrix} = \begin{bmatrix} -x_1x_3 \\ 2x_1x_3 \\ -x_1x_3 \end{bmatrix}. \tag{34}$$

If $x_1(0) = 1$, $x_2(0) = 1$, and $x_3(0) = 0$, then the exact solution is

$$\begin{bmatrix} x_1(t) \\ x_2(t) \\ x_3(t) \end{bmatrix} = \begin{bmatrix} \frac{1}{t+1} \\ \frac{2t}{t+1} \\ \frac{1}{t+1} \end{bmatrix}. \tag{35}$$

Note that for the exact solution, $\lim_{t\to\infty} x_1(t) = \lim_{t\to\infty} x_3(t) = 0$ and $\lim_{t\to\infty} x_2(t) = 2$. As one would expect from the chemical reaction equation, if there is no reverse reaction, all of species 1 binds with all of species 3 and produces species 2 at twice the concentration.

Now suppose a small numerical error occurs at $t = 999$, resulting in $x_1(999) = -0.001$ and $x_3(999) = -0.001$ instead of both being equal to positive 0.001. If the solution is computed accurately after the numerical error occurs, then

$$x_1(t) = x_3(t) = -\frac{1}{1999-t} \quad x_2(t) = 1.996 + \frac{2}{1999-t} \tag{36}$$

for $t > 999$. In other words, after the numerical error, x_1 and x_3 head increasingly rapidly toward negative infinity, driving x_2 increasingly rapidly toward positive infinity.

As the example demonstrates, the non-negative orthant property is fragile in the MAK case, a small numerical error at the wrong time can lead to extremely inaccurate simulation results. There are circuits that have a non-negative orthant property, but they do not suffer from the same fragility.

Circuits that satisfy the conditions at the beginning of Sect. 3.1 have a non-negative orthant property if the source currents are non-negative. Following the approach used to analyze the MAK case, we first give a "no hopping the fence" theorem.

Theorem 6. *If (17) is generated from a circuit that satisfies the conditions at the beginning of Sect. 3.1, and if* $\mathbf{v}(t), \mathbf{i}_s(t)$ *in (17) are in the non-negative orthant, then for all* j *such that* $\mathbf{v}_j(t) = 0$, $\frac{d}{dt}\mathbf{v}_j(t) \geq 0$.

In order to demonstrate that the non-negative orthant property of circuits is not fragile, we generalize Theorem 6, to a stronger "no hopping the fence" theorem.

Theorem 7. *If (17) is generated from a circuit that satisfies the conditions at the beginning of Sect. 3.1 and* $\mathbf{i}_s(t)$ *in (17) is in the non-negative orthant, and if, for all* i, $\mathbf{v}_i(t) \geq -\alpha$ *for some non-negative* α, *then for all* j *such that* $\mathbf{v}_j(t) = -\alpha$, $\frac{d}{dt}\mathbf{v}_j(t) \geq 0$.

Note that Theorem 7 is equivalent to Theorem 6 when $\alpha = 0$.

Passivity is the key property of our restricted class of circuits that leads to Theorem 7, though the positive diagonal form of the capacitance matrix also plays a role.

Proof. First note that $\mathbf{C}_{j,j}\frac{d}{dt}\mathbf{v}_j(t)$ is equal to the sum of currents flowing into node j minus the sum of currents leaving node j. By passivity, if $\mathbf{v}_j(t) = -\alpha$ and $\mathbf{v}_k(t) \geq -\alpha$ for all k, element currents incidence on node j are either flowing into node j or are zero. Note, this includes elements connected between node j and ground. If $-\alpha < 0$, these currents flow into node j, and if $-\alpha = 0$, these currents are zero. Then since $\mathbf{C}_{j,j} > 0$ by assumption, $\frac{d}{dt}\mathbf{v}_j(t) \geq 0$. □

Following a parallel argument to the one for Theorem 5, one can prove the following theorem.

Theorem 8. *If (17) is generated from a circuit that satisfies the conditions at the beginning of Sect. 3.1, and if the source currents are always non-negative, then for any trajectory* $\mathbf{v}(t)$, $t \geq 0$ *satisfying (17), if there is a non-negative* α *such that* $\mathbf{v}_j(0) \geq -\alpha$ *for all* j, *then* $\mathbf{v}_j(t) \geq -\alpha$ *for all* j *and for all* $t \geq 0$.

There is no circuit equivalent to the fragile example of the MAK case. If a numerical error results in computed voltages that are slightly more negative than $-\alpha$, from that point on, the node voltages will be restricted to a slightly larger range of values, and *no catastrophe will occur.*

4 Examples

In this last section, we present a few results that demonstrate the efficiency of exploiting the Kronecker product form for MAK when using scripting languages, and a second example demonstrating that conservation constraints must be considered to compute correctly oscillator parametric sensitivity.

4.1 Exploiting the Kronecker Form

Over the last decade, users of computational modeling have become progressively more willing to trade performance for flexibility. In most fields, users are moving away from specialized packages with predefined functionality, and moving toward scripting languages with built-in high performance graphics and core numerical algorithms (e.g. Mathematica, Matlab, Octave, Scipy-Pylab). Circuit simulation is an exception to this trend, as the analyses required are few and are well-understood, and there is an economy of scale due to the vast number of dedicated users. On the other hand, there are not nearly so many dedicated users of MAK models, and the analyses they require are a moving target [18–20].

In order to extract performance from a scripting language, it is important to use built-in functions as much as possible. In such a setting, the Kronecker-product based format for MAK-generated differential equation systems in (18) has a number of advantages. To see these advantages, consider combining (18) with (19) while also denoting an explicit dependence on a vector of parameters, **p**,

$$\frac{d}{dt}\mathbf{x}(t,\mathbf{p}) = \mathbf{f}(\mathbf{x}(t,\mathbf{p}),\mathbf{p}) + \mathbf{u}(t) = \mathbf{K}^{(0)}(\mathbf{p})\mathbf{x}(t,\mathbf{p}) + \mathbf{K}^{(1)}(\mathbf{p})\left(\mathbf{x}(t,\mathbf{p}) \otimes \mathbf{x}(t,\mathbf{p})\right) + \mathbf{u}(t), \tag{37}$$

where for simplicity, we assume $\mathbf{p} \in \mathbb{R}^{n_p}$ is a vector of reaction rate constants.

As MAK systems are typically stiff, with a mixture of time constants that range from milliseconds to hours, implicit time-integration schemes are far more efficient than explicit schemes for computing the solution to (37). Implicit schemes use Newton's method to solve the state-update equation at each time-step, and Newton's method usually requires the Jacobian. An advantage of the Kronecker-product form for MAK is that the Jacobian has an easily evaluated explicit form,

$$\frac{d\mathbf{f}}{d\mathbf{x}}(\mathbf{x}) = \mathbf{K}^{(0)} + \mathbf{K}^{(1)}\left(\mathbf{x} \otimes \mathbf{I} + \mathbf{I} \otimes \mathbf{x}\right). \tag{38}$$

Since many of the rate constants in MAK models are determined by fitting to measured data, it is often necessary to compute the parametric sensitivity of a species concentration. The standard forward method for computing such sensitivities is derived by differentiating (37) with respect to **p**, thus generating a sensitivity differential equation

$$\frac{d}{dt}\frac{d\mathbf{x}}{d\mathbf{p}}(t,\mathbf{p}) = \frac{d\mathbf{f}}{d\mathbf{x}}(\mathbf{x}(t,\mathbf{p}))\frac{d\mathbf{x}}{d\mathbf{p}}(t,\mathbf{p}) + \frac{d\mathbf{f}}{d\mathbf{p}}(\mathbf{x}(t,\mathbf{p})). \tag{39}$$

where $\frac{d\mathbf{f}}{d\mathbf{x}}(\mathbf{x}(t,\mathbf{p}))$ is given in (38), and the matrix $\frac{d\mathbf{f}}{d\mathbf{p}} \in \mathbb{R}^{n \times n_p}$ has a simple explicit formula if one considers each column individually. The ith column of $\frac{d\mathbf{f}}{d\mathbf{p}}$ is the sensitivity of $\mathbf{f}$ with respect to parameter $\mathbf{p}_i$,

$$\frac{d\mathbf{f}}{d\mathbf{p}_i}(\mathbf{x}(t,\mathbf{p})) = \frac{d\mathbf{K}^{(0)}}{d\mathbf{p}_i}\mathbf{x}(t,\mathbf{p}) + \frac{d\mathbf{K}^{(1)}}{d\mathbf{p}_i}(\mathbf{x}(t,\mathbf{p}) \otimes \mathbf{x}(t,\mathbf{p})). \tag{40}$$

Note that the matrices, $\frac{d\mathbf{K}^{(0)}}{d\mathbf{p}_i}$ and $\frac{d\mathbf{K}^{(1)}}{d\mathbf{p}_i}$ do *not* depend on $\mathbf{p}$, as both matrices are linear functions of the rate constants.

As (37), (38), and (40) demonstrate, using the Kronecker product form implies that evaluating quantities needed for time-integration, and for parametric sensitivity evaluation, become a combination of matrix-vector and Kronecker products. These products can be computed extremely efficiently, but only if the scripting language has very good sparse matrix support. For example, $\frac{d\mathbf{K}^{(1)}}{d\mathbf{p}_i} \in \mathbb{R}^{n \times n^2}$ is a huge sparse matrix, but could easily have as few as one nonzero entry.

We used Matlab [21] to create an easily extensible toolbox, KroneckerBio [18, 19], with a variety of analysis methods implemented using the Kronecker-product representation of MAK models. To demonstrate that such an approach can achieve reasonably good performance on large MAK models, we used two test biochemical models: a discretized partial differential equation (PDE) of antibody penetration into a tumor, and the enzymatic cascade (EC) consisting of an interacting chain of the modules studied in [22].

Our PDE model consists of a simple binding reaction where a mobile antibody binds to an immobilized tumor antigen. In this model, the unbound antibody is free to diffuse through the tumor. We model this process as a discretized PDE, noting that diffusion can be represented within the mass-action framework as a first-order reaction. We vary the number of compartments to generate models of different scale. The enzymatic cascade model is comprised of a chain of enzymatic push-pull loops, in which a substrate is modified to an active form by one enzyme and subsequently deactivated by another. The active substrate in each step is free to catalyze the activation of the substrate in the next level. As with the PDE example, we generate models of different scales by changing the length of the cascade.

To demonstrate that our implementation was capable of integrating large models, we implemented a series of models of increasing size. The results of this study are shown in Table 1. We find that simulation time grows less than quadratically for models of up to thousands of species. It should be noted that biochemical models comprising thousands of equations are rarely found in current literature.

Table 1 Simulation run times in seconds for models of varying size. The number of equations per model scales with the number of sub-models (e.g. PDE compartments or EC cycles) and with the number of equations per sub-model

Number of species	PDE ($N = 3$)	EC ($N = 5$)
N	0.0302 s	0.0560 s
$5N$	0.0332 s	0.1016 s
$10N$	0.0343 s	0.1185 s
$50N$	0.0834 s	0.7823 s
$100N$	0.2502 s	2.969 s
$500N$	4.8688 s	80.95 s
$1000N$	19.231 s	365.4 s

4.2 Oscillator Sensitivity Analysis

The numerical techniques for analyzing oscillators whose trajectories satisfy systems of autonomous ordinary differential equations, such as the systems generated by circuits or MAK models, have been investigated from so many perspectives it is hard to imagine that there is anything new to report. The basic finite-difference, spectral, and shooting-Newton methods have been known for decades [23, 24], and a number of research groups in different application domains have investigated approaches to computing parametric sensitivities [25–29].

Nevertheless, there are additional subtleties in the oscillator problem when there are conservation constraints. One such subtlety shows itself in the simplest of experiments. Consider the following differencing procedure for computing oscillator parametric sensitivity:

1. Starting from an initial condition, $\mathbf{x}(0)$, integrate the differential equation system until the computed trajectories form a periodic orbit with period T;
2. Perturb some system parameter of interest;
3. Integrate the perturbed differential equation system, again starting from the initial condition $\mathbf{x}(0)$, until the computed trajectories form a new periodic orbit with period $\hat{T}$.

From the analysis in (26), (27) and (28), we know that time-integration methods are unaffected by conservation constraints, so even if the above method is inefficient, it should be accurate. A far more efficient approach to computing oscillator parametric sensitivities is to differentiate the two-point boundary value problem associated with periodic steady-state, and methods based on such an approach have proven to be both efficient and accurate [27, 28]. However, if there are conservation constraints, these more efficient approaches *can lead to results that do not match the results from the above differencing algorithm.* Even the period parametric sensitivity may not be correct.

To demonstrate the difficulty, we consider an oscillating biochemical network harboring conservation constraints: a MAK model of the *in vitro* cyanobacterial circadian clock [30, 31]. This network consists of three proteins—KaiA, KaiB, and KaiC—that associate with one another and undergo modification. Surprisingly, the canonical behavior of a circadian clock, persistent oscillations with a period of about 24 hours, can be experimentally reconstituted in a minimal solution containing only the three purified proteins and their cofactors [32].

Our model of this network, based on that of [30], tracks every complex formed by the association and dissociation of the three Kai proteins, and consists of $n_x = 73$ state variables, $n_r = 277$ distinct chemical reactions, and $n_p = 22$ parameters. However, the combinatorial complexity of these reactions can be abstracted to yield a simple diagram containing 12 species representing the four phosphorylation states of KaiC (U—unphosphorylated; S—phosphorylated at Ser431; T—phosphorylated at Thr432; D—doubly phosphorylated) in three KaiA-related states (lower plane—free KaiC; middle plane—KaiC:KaiA complexes; upper plane—KaiC* after modification by KaiA) (Fig. 2).

Applying Theorem 2, we computed the left null space of the stoichiometric matrix associated with our circadian clock model, and uncovered three conservation constraints. We then computed the oscillator's periodic steady-state using a shooting-Newton method [24], both with and without explicit conservation constraints. To include explicit conservation constraints in the shooting-Newton method, we used an approach analogous to the one in (32).

The periods computed with and without the explicit conservation constraints were 20.47 and 20.00 hours, respectively. The period computed with the constraints was almost half an hour longer, but more importantly, it was also more accurate. That is, 20.47 hours was a much closer match to the period computed from the computationally-expensive brute-force approach of time-integrating the circadian clock oscillator for thousands of oscillation periods (the virtual equivalent of years).

After using the shooting-Newton method to compute the periodic steady-state, we used an approach similar to that in [27, 28] to compute the sensitivity of the oscillator period to the 22 model parameters. Again we computed the sensitivities

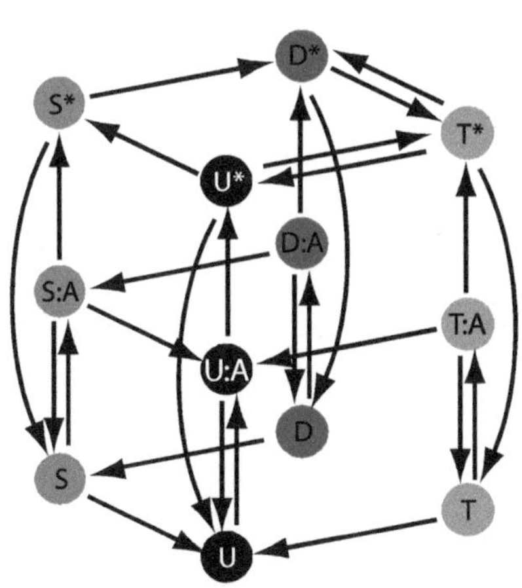

Fig. 2 The twelve major states of KaiC are pictured, with arrows representing reactions that modulate transfer between states

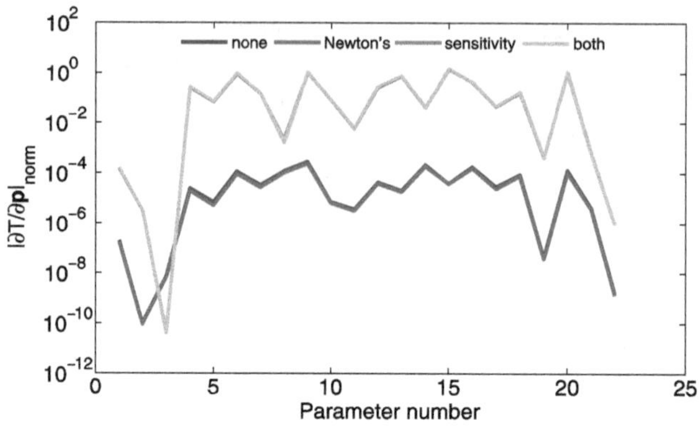

Fig. 3 Normalized parametric period sensitivity versus parameter index for the 22 parameters in the circadian clock model

with and without explicit conservation constraints, and again we added the constraints using an approach analogous to the one in (32) [31]. We then had four sets of 22 period sensitivities, for four scenarios:

1. completely ignoring the explicit conservation constraints,
2. using the explicit conservation constraints, but only for the shooting-Newton periodic steady-state computation,
3. using the explicit conservation constraints, but only when computing sensitivities, *NOT* when computing the steady-state,
4. using the explicit conservation constraints for both the periodic steady-state and sensitivity computations.

One would expect the four scenarios to result in four different sets of period sensitivities, but that is not quite the case. Consider the four plots of normalized period sensitivities versus parameter index in Fig. 3. In the figure, **none** denotes scenario 1, **Newton's** denotes scenario 2, **sensitivity** denotes scenario 3, and **both** denotes scenario 4. In Fig. 3, only two truly distinct curves are evident, corresponding to whether or not conservation constraints were included in the sensitivity calculation. These results indicate that, to achieve good accuracy, it is crucial to explicitly include conservation constraints in the sensitivity computation. Our results also suggest, somewhat surprisingly, that using an inaccurate periodic steady-state in the sensitivity computation has little impact, at least for this example.

5 Conclusions

The aim of this chapter was to examine the numerical implications of the differences between systems generated from circuits and from MAK, and we used both examples and theoretical analyses. In examining the issue of conservation constraints, an impractical issue in circuits but common in MAK, we demonstrated that such constraints must be handled explicitly if parametric sensitivities are to be computed accurately. We also examined the issue of how trajectories are kept in the non-negative orthant, and noted that even though species concentrations are non-physical if they are negative, small numerical errors can cause them drop below zero, often with catastrophic results. This fragility of the non-negative orthant is not replicated in circuits, at least for the setting we considered here. Finally, we demonstrated that by using Kronecker products to represent the differential equation systems generated by MAK, the resulting representation can be analyzed very efficiently using scripting languages like Matlab.

Acknowledgements The authors would like to thank Professor Peng Li for his extraordinarily careful proof-reading of drafts of this paper. This work was partially supported by the National Institutes of Health (CA112967), a Cancer Research Institute postdoctoral fellowship (J. Toettcher), the MIT Portugal Program, the Singapore MIT Alliance for Research and Technology, and the Singapore-MIT Alliance Computational Engineering and Computation and Systems Biology programs.

References

1. A. Kremling and J. Saez-Rodriguez, "Systems biology—An engineering perspective," *J. Biotechnol.*, vol. 129, pp. 329–351, 2007.
2. H. Kitano, "Systems biology: A brief overview," *Science*, vol. 295, pp. 1662–1664, 2002.
3. B.O. Palsson, *Systems Biology*, New York, NY, Cambridge University Press, 2006.
4. J.L. Wyatt, "Network representation of reaction-diffusion systems far from equilibrium," *Comput. Prog. Biomed.*, vol. 8, pp. 180–195, 1978.
5. F. Najm, *Circuit Simulation*, Hoboken, NJ, Wiley and Sons, 2010.
6. S. Agarwal and J. Roychowdhury, "Efficient multiscale simulation of circadian rhythms using automated phase macromodelling techniques," *Proc. Pacific Symp. Biocomputing*, vol. 13, pp. 402–413, 2008.
7. L.O. Chua, C.A. Desoer, and E.S. Kuh, *Linear and Nonlinear Circuits*, New York, NY, McGraw-Hill, 1987.
8. W.M. Deen, *Analysis of Transport Phenomena*, New York, NY, Oxford University Press, 1998.
9. A. Graham, *Kronecker Products and Matrix Calculus with Applications*, New York, NY, Halsted Press, 1981.
10. S. Mandal and R. Sarpeshkar, "Log-domain circuit models of chemical reactions," *Proceedings of the IEEE International Symposium on Circuits and Systems*, pp. 2697–2700, 2009.
11. H. Schichman and D.A. Hodges, "Modelling and simulation of insulated-gate field-effect transistor switching circuits," *IEEE Journal of Solid-State Circuits*, vol. SC-3, pp. 285–328, 1986.
12. L.J. Trajkovid, R.C. Melville, and S.C. Fang, "Passivity and nogain properties establish global convergence of a homotopy method for DC operating points," *Proc. IEEE Int. Symp. Circuits and Systems*, pp. 914–917, 1990.
13. H. de Jong, "Modeling and simulation of genetic regulatory systems: A literature review," *Journal of Computational Biology*, vol. 9, no. 1, pp. 67–103, 2002.
14. J. Stoer and R. Bulirsch, *Introduction to Numerical Analysis*, Third Edition, Springer, 2002
15. G.H. Golub and C.F. Van Loan, *Matrix Computations*, Baltimore, MD, Johns Hopkins Univ. Press, 1983.

16. I.W. Sandberg, "A nonnegativity-preservation property associated with certain systems of nonlinear differential equations," *Proc. IEEE Int. Conf. on Systems, Man and Cybernetics*, pp. 230–233, 1974.
17. J.L. Wyatt Jr., "Monotone sensitivity of nonlinear nonuniform RC transmission lines with application to timing analysis of digital MOS integrated circuits," *IEEE Trans. Circuits Syst.*, CAS-32, pp. 28–33, 1985.
18. J.F. Apgar, J.E. Toettcher, D. Endy, F.M. White, and B. Tidor, "Stimulus design for model selection and validation in cell signaling," *PLoS Comput. Biol.*, vol. 4, no. 2, 2008.
19. J.F. Apgar, D.K. Witmer, F.M. White, and B. Tidor, "Sloppy models, parameter uncertainty, and the role of experimental design," *Mol. Bio. Syst.*, vol. 6, pp. 1890–1900, 2010.
20. J.E. Toettcher, C. Mock, E. Batchelor, A. Loewer, and G. Lahav, "A synthetic–natural hybrid oscillator in human cells," *PNAS*, vol. 107, no. 39, pp. 17047–17052, 2010.
21. Mathworks, *Matlab (tm) 2007a*. Natick, MA.
22. A. Goldbeter and D.E. Koshland, Jr., "An amplified sensitivity arising from covalent modification in biological systems," *Proc. Natl. Acad. Sci. USA*, vol. 78, pp. 6840–6844, 1981.
23. T.J. Aprille and T.N. Trick, "A computer algorithm to determine the steady-state response of nonlinear oscillators," *IEEE Trans. Circuit Theory*, vol. CT-19, pp. 354–360, 1972.
24. K.S. Kundert, J.K. White, and A. Sangiovanni-Vincentelli, *Steady-State Methods for Simulating Analog and Microwave Circuits*, Amsterdam, The Netherlands, Kluwer, 1990.
25. D.E. Zak, J. Stelling, and F.J. Doyle, "Sensitivity analysis of oscillatory (bio)chemical systems," *Comput. Chem. Eng.*, vol. 29, no. 3, pp. 663–673, 2005.
26. S.R. Taylor, R. Gunawan, L.R. Petzold, and F.J. Doyle, "Sensitivity measures for oscillating systems: Application to mammalian circadian gene network," *IEEE Transactions on Automatic Control*, vol. 53, pp. 177–188, 2008.
27. A.K. Wilkins, P.I. Barton, and B. Tidor, "The Per2 negative feedback loop sets the period in the mammalian circadian clock mechanism," *PLoS Comput. Biol.*, vol. 3, no. 12, 2007.
28. I. Vytyaz, D.C. Lee, P.K. Hanumolu, Un-Ku Moon, and K. Mayaram, "Sensitivity Analysis for Oscillators," *IEEE Transactions on Computer-Aided Design of Integrated Circuits and Systems*, vol. 27, no. 9, pp. 1521–1534, 2008.
29. A. Demir, "Floquet theory and non-linear perturbation analysis for oscillators with differential-algebraic equations," *International Journal of Circuit Theory and Applications*, vol. 28, no. 2, pp. 163–185, 2000.
30. M.J. Rust, J.S. Markson, W.S. Lane, D.S. Fisher, and E.K. O'Shea, "Ordered phosphorylation governs oscillation of a three-protein circadian clock," *Science*, vol. 318, no. 5851, pp. 809–812, 2007.
31. J.E. Toettcher, A.R. Castillo, G. Lahav, J.K. White, and B. Tidor, *in preparation.*
32. M. Nakajima, K. Imai, H. Ito, T. Nishiwaki, Y. Murayama, H. Iwasaki, T. Oyama, and T. Kondo. "Reconstitution of circadian oscillation of cyanobacterial KaiC phosphorylation in vitro," *Science*, vol. 308, no. 5720, pp. 414–415, 2005.

Circuit-Based Models of Biomolecular System Dynamics

Elebeoba E. May

Abstract Methods from the field of electrical engineering have been used for the modeling and analysis of biological systems. In this work we exploit parallels between electrical and biological circuits for simulation of biomolecular processes and systems. We review the development of BioXyce, an electrical circuit-based systems biology simulation platform, and demonstrate its use in simulating intracellular biochemical pathways. We present simulation results for metabolic pathways and eukaryotic signal transduction pathways important in host immune response.

1 Capturing the Dynamics of Living Circuits

The field of systems biology provides a framework for exploring how subcomponents of a biological system interact to cooperatively produce functional behavior. Continued advancements in the *-omics* fields, including genomics, proteomics, and metabolomics are providing new insight into molecular systems and mechanisms that integrate to form the biological pathways or the circuitry of living systems. Systems biology encompasses many different length and time scales, and involves the application of empirical and quantitative methods in the study of biological processes. This chapter will focus on quantitative molecular systems biology, and specifically address biological pathways or networks within prokaryotic and eukaryotic cells. We use quantitative methods and tools derived from electronic circuit analysis to model and simulate the functional behavior of biological pathways or circuits of interest.

Biological networks like electrical circuit systems are composed of large numbers of subsystems and interacting subcircuits. While electronic systems are composed of interconnected devices and components, a eukaryotic cell has membrane

Elebeoba E. May
Sandia National Laboratories, P.O. Box 5800, Albuquerque, NM, USA
e-mail: eemay@sandia.gov

P. Li et al. (eds.), *Simulation and Verification of Electronic and Biological Systems*,
DOI 10.1007/978-94-007-0149-6_7,

bound organelles that functionally localize cellular components. For example, Fig. 1 provides a simplified diagram of the cell that depicts three major functional areas of a cell: the cell membrane, the cytoplasm, and the nucleus [1].

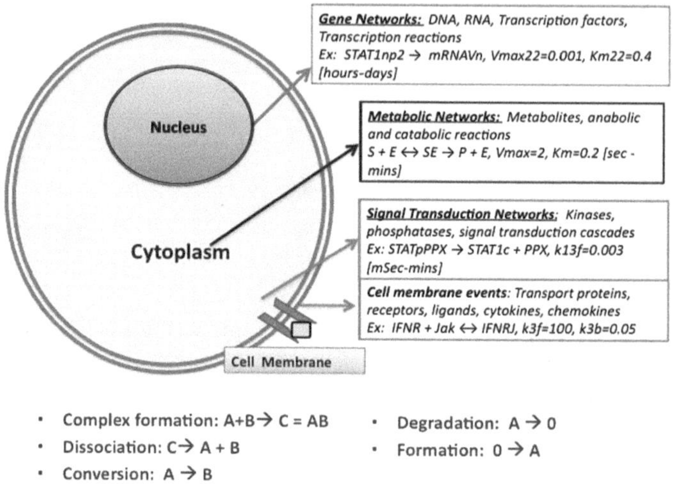

Fig. 1 Simplified diagram of a cell depicting three functional areas and example reactions and protein components associated with the region

The cell membrane helps separate the intracellular and extracellular space and contains embedded proteins or receptors that enable chemical signals that originate outside of a cell to be received and functionally propagated within the cell. Signal transduction pathways can be initiated by the binding of a chemical signal or ligand to a cell membrane receptor, which then leads to the activation of proteins in the cytoplasm. The cytoplasm, the intracellular space, contains proteins and protein complexes that carry out processes required for cell function, including metabolic processes and translation of messenger RNA to protein products. In eukaryotic cells (e.g. animal cell) the cytoplasm contains membrane bound organelles where specialized processes occur, such as the generation of ATP in the mitochondria. In general, prokaryotic cells (e.g. bacterial cell) do not have membrane bound organelles. While eukaryotes have a membrane bound nucleus as depicted, prokaryotic genomic material aggregates into a nucleus-like region called a nucleoid [1]. Genetic networks responsible for replication of genetic material and regulation of genes into messenger RNA transcripts are found within the nucleus. In the sections that follow, we will discuss and demonstrate how we use a circuit-based simulation platform to model the biological circuits involved in metabolic pathways, genetic networks, and signal transduction cascades.

1.1 BioXyce: An Electrical Circuit-Based Systems Biology Simulation Platform

Electrical systems, like biological systems, are made up of large numbers of components and interacting subnetworks. Simulation of these systems must produce accurate results in an efficient manner. Parallel circuit simulation tools, such as Xyce (http://www.cs.sandia.gov/xyce/) allow engineers to model and test very large-scale systems. The theoretical framework used in developing Xyce also provides the necessary tools for constructing a biological circuit simulator that can model multivariate, multiscale, hybrid biological networks.

We recently described the development and use of BioXyce (Fig. 2), an electronic circuit-based biological system simulation platform based on Xyce, to model and simulate single and multicellular biological networks [2]. At the cellular level,

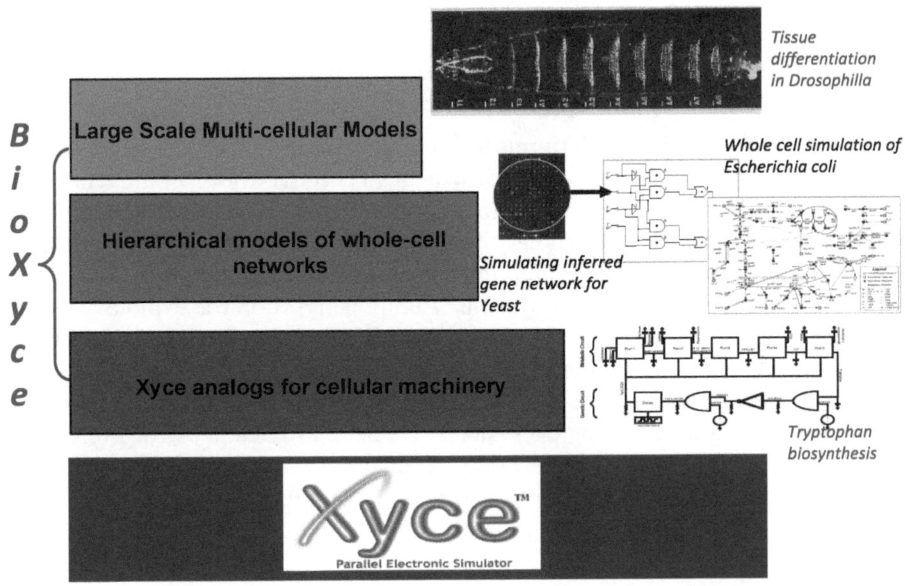

Fig. 2 BioXyce is a large-scale biological circuit simulation platform based on Xyce, a large-scale electronic circuit simulator

biochemical pathways are modeled as electrical circuits where signals are produced, propagated and consumed. BioXyce uses the following equivalents: chemical mass as charge, mass flux as electric current, concentration as voltage, stoichiometric conservation as Kirchhoff's voltage law, and mass conservation as Kirchhoff's current law.

Although several researchers have made and used parallels between biological networks and electrical systems in modeling and simulation [3–5], few exploit large-scale simulation and analysis tools used by electrical engineers in the simulation of biological networks. Similar to the abstractions used in flux balance analysis (FBA), the flow of metabolic and genetic substrates is synonymous to the flow of current through an electrical circuit [6]. In BioXyce metabolic reactions are modeled using analog subcircuits and analog behavioral models (ABM), where metabolite accumulation and degradation are represented using capacitors and resistors, respectively. Reactions in genetic regulatory circuits are represented using both CMOS-based logic sub-circuits that implement Boolean functions and Boolean kinetics implemented using ABM components. The enzymatic products of genetic pathways can be functionally modeled as "metabolic clocks" that control when metabolic reactions occur; when the enzyme level reaches a pre-defined threshold, the associated reaction occurs. With BioXyce, we can focus on subsystems like those described in this work as well as simulate large control networks consisting of entire cells, homogeneous cell cultures, or heterogeneous interacting host-pathogen networks in order to understand the dynamics and stability of such systems.

1.2 Modeling Biological Pathways Using BioXyce

Systems biology simulation platforms like E-Cell, Virtual Cell, COPASI, ChemCell, Cellerator, and Hy3S offer a variety of mathematical frameworks for modeling biochemical processes [7–12]. While there are similarities between BioXyce and these simulation platforms, the underlying circuit simulation platform on which BioXyce is constructed, is massively parallel; this enables BioXyce to readily model large-scale networks and multicellular systems. Additionally, BioXyce exploits the similarity between biological networks and electrical circuits to establish an electrical circuit-based paradigm for representing and simulating biological processes. This enables the use of advancements in the field of electronic circuit simulation and analysis for the study of biological systems. We demonstrate the use of BioXyce to model three types of biological networks: metabolic networks, gene networks, and signal transduction pathways. The following types of processes are modeled:

- Complex formation: $A+B \rightarrow AB$.
- Dissociation: $C \rightarrow A+B$.
- Conversion: $A \rightarrow B$.
- Degradation: $A \rightarrow 0$.
- Formation: $0 \rightarrow A$.

Using public domain resources (Table 1), we formulate our model as outlined:

1. Determine the initial network topology or stoichiometry of the system using KEGG (Kyoto Encyclopedia of Genes and Genomes), MetaCyc/BioCyc, or literature curation.
2. Create a set of ordinary differential equations that correspond to system stoichiometry and capture the dynamics of the network. Two general types of reactions illustrated in this chapter for $S \rightarrow P$ are:

Table 1 Public domain systems biology resources

Cellular Pathways
KEGG: http://www.genome.jp/kegg/
BioCyc: http://biocyc.org/
MetaCyc: http://metacyc.org/
Reactome: http://www.reactome.org/
BioCarta: http://www.biocarta.com/genes/index.asp
ExPASy Biochemical Pathways: http://www.expasy.org/cgi-bin/search-biochem-index
Empirical Data Repositories
Gene Expression Omnibus (GEO): http://www.ncbi.nlm.nih.gov/geo/
Reference Database of Immune Cells (RefDIC): http://refdic.rcai.riken.jp
Systems Immunology: http://systemsimmunology.org
Biochemical Reaction Kinetics
SABIORK: http://sabio.villa-bosch.de/SABIORK/
BRENDA: http://www.brenda.uni-koeln.de/

Mass action kinetics:

$$\frac{dP}{dt} = k \times [S]$$

where k is the reaction rate.

Michaelis-Menten kinetics:

$$\frac{dP}{dt} = \frac{V_{\text{Max}} \times [S]}{(K_M + [S])} = \frac{K_{\text{cat}} \times [E] \times [S]}{(K_M + [S])}$$

where V_{Max} is the maximum reaction velocity, K_M is the Michaelis-Menten constant, K_{cat} is the turnover rate, and E is the total enzyme concentration.

3. Using literature curation and databases like BRENDA (Braunschweig Enzyme Database), determine and incorporate initial values for kinetic rate parameters and initial concentration values.
4. Determine and refine parameter values for the system using empirical data, parameter estimation and optimization methods.

We have developed tools to automate portions of this process, including tools to automate the extraction of kinetic parameters from BRENDA and tools that generate the Xyce compatible netlist from a flat file stoichiometric description of the system. The netlist is executed using Xyce and time-varying voltage values returned by Xyce correspond to the dynamic concentration values. In the following sections we demonstrate the use of BioXyce for modeling metabolic networks (Sect. 2), gene networks (Sect. 3), and signal transduction pathways (Sect. 4).

2 Simulating Metabolic Processes with Genetic Control

Using *BioXyce*, we model and simulate the metabolic and genetic regulation of tryptophan biosynthesis. In this pathway the genetic and metabolic networks directly interact. The final output of the metabolic circuit is tryptophan, the key input into the gene regulatory network that coordinates the expression of enzymes that catalyze the metabolic reactions. In this work we refer to these systems as hybrid models since we integrate two types of biological networks into a single model.

As discussed in [2], metabolic models described below use the Xyce analog behavioral modeling element, the *BSOURCE*, to model each enzymatic reaction in general as:

$$Rate_P = C_P \times fRate \times (S_1^{C_1}) \times (S_2^{C_2}) \times \ldots (S_N^{C_N}) \quad (1)$$

where $Rate_P$ is the rate of production or consumption of metabolite P and C_P is the stoichiometric coefficient of metabolite P; $fRate$ is the metabolic reaction rate and would be derived from the Michaelis-Menten rate $V_{\max}\frac{[S]}{K_M+[S]}$. S_I is the concentration of the Ith metabolic substrate and C_I is the corresponding stoichiometric coefficient, where $I = 1\ldots N$. To connect the gene network and the metabolic reaction network, we developed CMOS-based transmission and logic gates and incorporated the gates into the stoichiometric reaction subcircuit. This permits the output of the gene network to regulate when and if a metabolic reaction occurs. For small-scale simulations, the use of CMOS-based logic gates to model gene regulatory networks was sufficient, however for larger gene networks we developed Boolean kinetics-based logic gates, which is described in Sect. 3.

2.1 Tryptophan Biosynthesis

In bacteria, the synthesis of tryptophan is regulated primarily by feedback inhibition where the presence of tryptophan negates the production of four enzymes that catalyze various steps in tryptophan biosynthesis: anthranilate synthase, anthranilate phophoriosyl transferase, indole glycerol phosphate synthase, and tryptophan synthase [13]. The presence of tryptophan represses the transcription of the tryptophan operon. The operon contains five structural genes: trpE, trpD, trpC, trpB, and trpA. The repressor gene, trpR, the promoter region, the operator region and the trpL leader gene are also part of the tryptophan operon. Figure 3 shows the metabolic pathway and tryptophan operon genes associated with each biosynthetic step.

In the absence of tryptophan (or very low levels of tryptophan in the bacterial cell), the RNA polymerase molecule is able to transcribe the region of the DNA that codes for the trpEDCBA genes. The ribosome translates the resulting messenger RNA (mRNA) and produces the enzymes that catalyze the metabolic reactions for tryptophan biosynthesis. Once a significant level of tryptophan is present in the cell the tryptophan feeds back into the genetic regulation processes and repression occurs in two steps. Tryptophan, acting as a corepressor, binds to the inactive re-

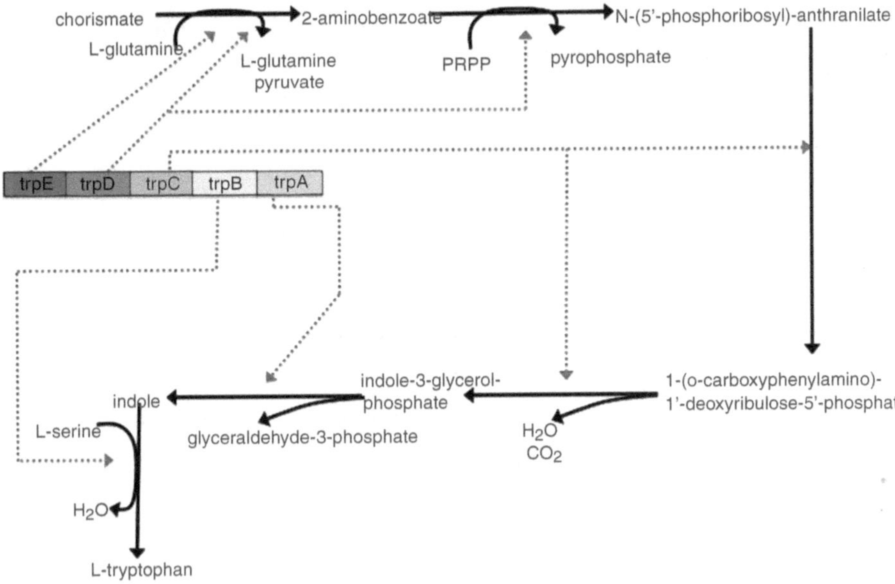

Fig. 3 Genetic and metabolic pathways for the activation of tryptophan biosynthesis. *Derived from pathway outlined in BioCyc.org and http://www.med.sc.edu:85/mayer/genreg7.jpg*

pressor protein (an aporepressor), which is the result of transcribing and translating gene trpR. The activated repressor/tryptophan complex, a holorepressor, binds to the tryptophan operon. This prevents the RNA polymerase from transcribing the trpEDCBA gene, hence the enzymes needed for catalysis of the metabolic reactions in the biosynthesis pathway are not formed and tryptophan biosynthesis stops.

The genetic regulation, derived from Xiu et al.'s work [13], is modeled using Boolean circuits with tryptophan concentration as the only variable input and the presence (logic 1) or absence (logic 0) of trpEDCBA as the output. The current implementation leaves room for the incorporation of a more detailed genetic circuit model that takes into account factors such as mRNA concentration, genetic rate constants, and gene levels. The existing model makes the simplifying assumption that the presence of the trpEDCBA gene correlates to the successful production of the corresponding enzymes.

The stoichiometric reactions involved in tryptophan biosynthesis and the associated enzymes are:

$$\begin{array}{rl} \text{chor} + \text{gln} \rightarrow \text{glu} + \text{pyr} + \text{an} & \text{(anthranilate synthase; trpE, trpD)} \\ \text{an} + \text{prpp} \rightarrow \text{ppi} + \text{npran} & \text{(anthranilate phosphoribosyl transferase; trpD)} \\ \text{npran} \rightarrow \text{cpad5p} & \text{(phosphoribosylanthranilate isomerase; trpC)} \\ \text{cpad5p} \rightarrow \text{co2} + \text{igp} & \text{(indole-3-glycerol phosphate synthase; trpC)} \\ \text{igp} + \text{ser} \rightarrow \text{t3p1} + \text{trp} & \text{(tryptophan synthase, beta subunit dimer; trpB)} \end{array} \quad (2)$$

Table 2 lists the abbreviations for the metabolites involved and modeled in our BioXyce simulation of tryptophan biosynthesis.

Table 2 Abbreviations for metabolites involved in tryptophan biosynthesis

Abbreviation	Metabolite
chor	Chorismate
gln	Glutamine
glu	Glutamate
pyr	Pyruvate
an	Antranilate
prpp	Phosphoribosyl pyrophosphate
ppi	Pyrophosphate
npran	N-(5'-phosphoribosyl)-anthranilate
cpad5p	1-(O-Carboxyphenylamino)-1'-deoxyribulose-5'phosphate
co2	Carbon dioxide
igp	Indole glycerol phosphate
ser	Serine
t3p1	Glyceraldehyde 3-phosphate
trp	Tryptophan

2.2 *Simulating the Tryptophan Hybrid Network*

Figure 4 depicts the hybrid metabolic and genetic tryptophan circuit. Stoichiometric data (from *http://gcrg.ucsd.edu/* and *http://biocyc.org/*) and qualitative descriptions available in literature [13, 14] are used to construct the coupled metabolic and genetic circuit netlist below:

```
************************************************************************
Tryptophan Biosynthesis Circuit

**************** Genetic Circuit
* Initial conditions for logic circuit
** Set voltage and clock values
Vdddev nVdd 0 5V
VddNdev nVddN 0 -5V
Vset set 0 0V
Vreset reset 0 0V

Vclk genClk 0 PULSE(-5V 5V 0 0.1 0.1 .8 1.6)
xNeg_gclk genClk genClkNot neg

* Initial Conditions:
*     [apoRep] = 10 M  10V=positive 1 in CMOS logic gates
*     [mRNA] = 10 M
Vin_apoRep apoRep 0 5V
Vin_mRNA mRNA 0 5V
```

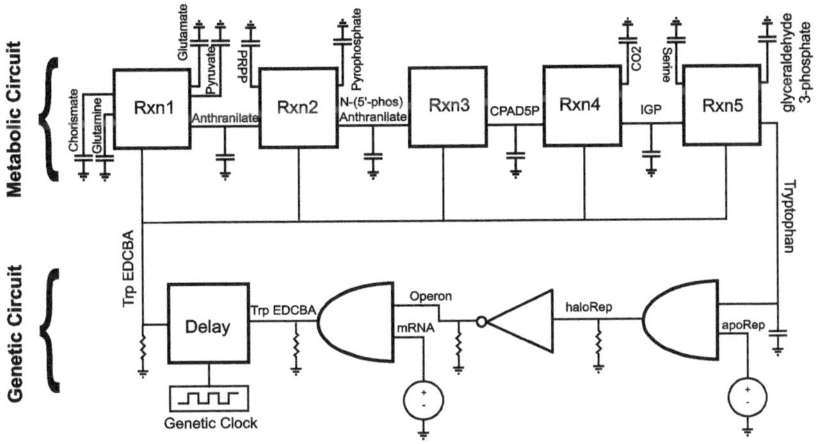

Fig. 4 Circuit diagram of tryptophan genetic and metabolic network

```
xAnd_holoRep trp apoRep holoRep nVdd and2
xNot_opr holoRep opr nVdd not
xAnd_trpEDCBAprev opr mRNA trpEDCBAprev nVdd and2
xDff_trpEDCBA trpEDCBAprev enzymeIn enzymeInNot nVdd
    nVddN genClk genClkNot set reset dff
xEClk_trpEDCBA enzymeIn trpEDCBA trpEDCBANot enzymeClk
        PARAMS: oldmaxVal=5 newmaxVal=5

RresHoloRep holoRep 0 100K
RresOpr opr 0 100K
RresTrp trp 0 100K
RresTrpEDCBAprev trpEDCBAprev 0 100K

*********************** Metabolic Circuit
* Initial Conditions for metabolic network
*     [chor] = 0.01 M
*     [gln] = 0.01 M
*     [prpp] = 0.01 M
*     [ser] = 0.01 M
*
* Externally provided
CcapChor chor 0 1 IC=5
CcapGln gln 0 1 IC=5
CcapPrpp prpp 0 1 IC=5
CcapSer ser 0 1 IC=5

*   Internally produced but not consumed
CcapGlu glu 0 1 IC=.01
CcapPyr pyr 0 1 IC=.01
CcapPpi ppi 0 1 IC=.01
CcapCo2 co2 0 1 IC=.01
CcapT3p1 t3p1 0 1 IC=.01
```

```
* Internally produced and consumed in tryp operon genetic network
CcapTrp trp 0 1 IC=.01

* Internally produced and consumed in tryp metabolic network
CcapAn an 0 1 IC=.01
CcapNpran npran 0 1 IC=.01
CcapCpad5p cpad5p 0 1 IC=.01
CcapIgp igp 0 1 IC=.01

** chor + gln -> glu + pyr + an
xRxn_ChorGln_GluPyrAn chor gln glu pyr an trpEDCBANot trpEDCBA
        nVdd nVddN Rxn2To3
        PARAMS: r1Stio=1 r2Stio=1 p1Stio=1 p2Stio=1 p3Stio=1 fRate=1
** an + prpp -> ppi + npran
xRxn_AnPrpp_PpiNpran an prpp ppi npran trpEDCBANot trpEDCBA
        nVdd nVddN Rxn2To2
        PARAMS: r1Stio=1 r2Stio=1 p1Stio=1 p2Stio=1 fRate=1
** npran -> cpad5p
xRxn_Npran_Cpad5p npran cpad5p trpEDCBANot trpEDCBA
        nVdd nVddN Rxn1To1
        PARAMS: r1Stio=1 p1Stio=1 fRate=1
** cpad5p -> co2 + igp
xRxn_Cpad5p_Co2Igp cpad5p co2 igp trpEDCBANot trpEDCBA
        nVdd nVddN Rxn1To2
        PARAMS: r1Stio=1 p1Stio=1 p2Stio=1 fRate=1
** igp + ser -> t3p1 + trp
xRxn_IgpSer_T3p1Trp igp ser t3p1 trp trpEDCBANot trpEDCBA
        nVdd nVddN Rxn2To2
        PARAMS: r1Stio=1 r2Stio=1 p1Stio=1 p2Stio=1 fRate=1

** Analysis
.TRAN 0.25s 20s 0
** Output
.PRINT TRAN V(genClk) V(holoRep) V(trpEDCBAprev) V(enzymeIn)
        V(trpEDCBA) V(chor) V(an) V(glu) V(npran) V(cpad5p)
        V(igp) V(trp)
*********************************************************************
```

The netlist was executed using Xyce and Fig. 5 shows the results of the tryptophan circuit simulation. In Fig. 5 the horizontal axis is simulation time (tSim) and the vertical axis is concentration of the enzyme and metabolites involved. The simulation results illustrate features of the well known tryptophan regulation process. At simulation time tSim = 0, tryptophan is not present in the system so the genetic circuit permits the transcription (and ultimately translation) of the trpEDCBA enzyme (first graph of Fig. 5) which catalyzes reactions in the tryptophan biosynthesis pathway. As depicted in Fig. 5, tryptophan (TRP; fifth graph) is produced as chorismate (CHOR; second graph) and 1-(o-carboxyphenylamino)-1'-deoxyribulose-5'-P (CPAD5P, fourth graph) are consumed. While CPAD5P is produced and consumed during tryptophan biosynthesis, Glutamate (GLU; third graph) is produced but not consumed by any step in the biosynthesis pathway. Once a significant level of tryptophan is present in the system (e.g. simulation time tSim = 4) transcription of the trpEDCBA enzyme is inhibited. As tryptophan is degraded (tSim = 4 to

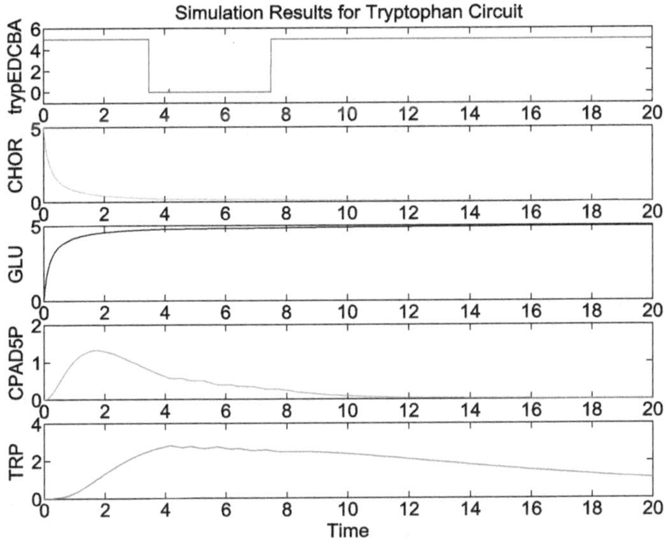

Fig. 5 Simulation results for the tryptophan genetic and metabolic circuit

tSim = 7.5) transcription of the trpEDCBA gene resumes once tryptophan levels fall below threshold (currently the boolean logic sub-circuit is activated when levels fall below approximately 0.5*(Maximum Substrate Concentration)). After tSim = 7.5, although tryptophan levels increase slightly, there are not enough reactants in the system to produce sufficient tryptophan levels in the presence of the degradation element.

3 Boolean Kinetics Framework for Simulating Large Scale Gene Networks

Quantitative descriptions of biochemical pathways that govern the functional behavior of cells are constructed using computational frameworks such as mathematical logic and Boolean algebra, differential equations, and graph theory [15–18]. The question the model attempts to answer determines the mathematical framework used. For instance, models used in the analysis of metabolic pathways typically employ differential equations in order to recover parameters useful in predicting growth and by-product secretion rates. But when the model is used to determine the qualitative behavior of a pathway, such as gene regulatory pathways, a logical representation of the system may be a necessary and sufficient simplification. In this section we use a qualitative approach to model a whole-cell gene network, employing Boolean algebra as the framework for constructing our model.

3.1 Modeling Gene Networks with Boolean Logic

The field of Computer Engineering uses mathematical logic in the design of digital systems for computer hardware. There are several ways to represent a system using mathematical logic (permits us to categorically evaluate the truth of a set of statements) and Boolean algebra (algebraic method for combining logic values). Possible system representation includes [19]:

- Truth table: a list of all possible input combinations and the resulting output values for the given system.
- Logic gates (schematic representation): implementation of the system using interconnected primitive components (logic gates, ex: OR, AND, NOT).
- Boolean equations: an algebraic short-hand for representing the truth table of a function or system.

The use of mathematical logic to quantify biological processes began in the late nineteen sixties with Kaufman's work which used Boolean networks to analyze biochemical pathways and interactions of immune molecules [20]. Boolean networks model each node or gene in the case of gene networks as on (a value of 1) or off (a value of 0). The updating function for each node is dependent on the current state of all its k input nodes. Although Boolean networks are a gross approximation of the intricate interactions present in biochemical networks or immune response networks, they have yielded useful results in the analysis or experimental gene expression data and for reverse engineering of biological networks [21].

To accommodate whole organism gene networks, we developed logic gates using Boolean kinetics. AND, OR, and NOT gates are simulated using a variation of the dynamic boolean equations used by Shen-Orr, et al., which in turn is based on the Boolean kinetics equation presented in McAdams and Arkin [18, 22]. The general equation for a gene Y whose output concentration is dependent on the concentration of an input gene X:

$$\frac{dY}{dt} = F(X, T_Y) - aY \tag{3}$$

where in our implementation $F(X, T_Y)$ is a step function dependent on the activation threshold value, T_Y. The constant a is related to the degradation rate of Y. Variations of the general form, (3), are used to represent the logic gates as depicted in Table 3:

Table 3 Boolean kinetics equations for implementation of genetic logic gates

Logic relation	Boolean kinetics equation
NOT (X)	$\frac{dY}{dt} = F(V_{\text{Max}} - X, T_Y) - aY$
X_1 AND X_2	$\frac{dY}{dt} = F(X_1, T_{Y,1}) * F(X_2, T_{Y,2}) - aY$
X_1 OR X_2	$\frac{dY}{dt} = F((F(X_1, T_{Y,1}) + F(X_2, T_{Y,2})), T_{Y,OR}) - aY$

Below is the netlist implementation of the basic single input and double input Boolean kinetics logic gates represented in Table 3:

```
************** Boolean Kinetics Logic Gates*************************

*********************************************************************

***** NOT Subcircuit
.SUBCKT NOT in1 outF PARAMS: koutF=1
***SUBCIRCUIT NOT - Dynamic Boolean NOT gate
*** Equation:  outF = NOT in1=> d outF/dt = F(Vmax - in1,TNot)- kout*outF
*** Note:  Vmax=Max voltage, TNot=0.5
.PARAM vMax=1
.PARAM TNot=0.5
BoutF outF 0 V={SDT(U(vMax-V(in1) - TNot) - V(outF)*koutF)}
.ENDS
*********************************************************************

***** AND Subcircuit
.SUBCKT AND in1 in2 outF PARAMS: T1=0.5 T2=0.5 koutF=1
*** SUBCIRCUIT AND - Dynamic Boolean AND gate
*** Equation:  outF = in1 AND in2 => d outF/dt = F(in1,T1)*F(in2,T2)
                 - kout*outF
BoutF outF 0 V={SDT(U(V(in1) - T1)*U(V(in2) - T2)-V(outF)*koutF)}
.ENDS
*********************************************************************

***** OR Subcircuit
.SUBCKT OR in1 in2 outF PARAMS: T1=0.5 T2=0.5 koutF=1
*** SUBCIRCUIT OR - Dynamic Boolean OR gate
*** Equation:  outF = in1 AND in2 => d outF/dt = F[(F(in1,T1) +
                 F(in2,T2)),TOR] - kout*outF
*** Note:  TOR=0.5
.PARAM TOR=0.5
BoutF outF 0 V={SDT(U((U(V(in1) - T1) + U(V(in2) - T2)) - TOR)
             - V(outF)*koutF)}
.ENDS
*********************************************************************
```

3.2 Using a Boolean Kinetics Model of Gene Regulation in the Tryptophan Hybrid Network

Genetic interactions in the tryptophan circuit were initially represented using CMOS-based logic components. We found that as the network size increased (discovered for larger, eighty-one node gene network simulation), successful simulation became prohibitive. Therefore we devised non-CMOS based logic gates using the Boolean kinetics method described [18, 22]. The Boolean kinetics representation allowed us to produce analog waveforms for the gene network and lends towards a more compatible representation of gene and enzyme levels for hybrid networks. As Fig. 6 shows, the use of the Boolean kinetics framework for genetic circuits produced comparable results for tryptophan biosynthesis. The genetic subcircuit appears to more

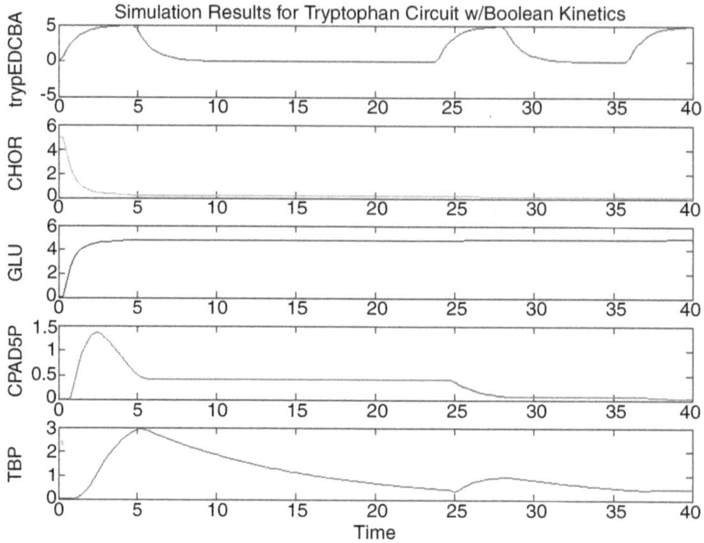

Fig. 6 Simulation results of tryptophan genetic and metabolic circuit using a Boolean kinetics framework for gene network modeling

rapidly impact the behavior of the metabolic circuit given a change in *trpEDCBA*. The output of the gene network are analog waveforms with smoother transients. Additionally the model responds more rapidly to changes in tryptophan levels, producing the needed genes to drive the biosynthetic pathway.

3.3 Simulation of Whole-Cell Inferred Gene Network

We used the Boolean kinetics framework to simulate the complete inferred gene network of yeast. Time series microarray data for the yeast system was clustered into gene groups, which are referred to as meta-genes. The time-dependent expression profile of each meta-gene was discretized and served as the input to the genetic network inference algorithm [23]. The algorithm produced a truth table that includes every possible input/output relationship for the node. Each meta-gene node, Y, has 2^k entries, where k is the number of meta-gene nodes that influence the output of Y. As the k increases the size of the truth table grows exponentially and the number of elements in the circuit grows substantially. Since the truth table representation is exhaustive and most likely redundant, simulation of the complete truth table would be excessive. We applied logic reduction rules from Boolean algebra to produced a minimized representation of the inferred gene network. The minimized form contains all of the behavioral relationships and leads to a simpler circuit realization.

We performed a two-level Boolean minimization on the truth table representation of the inferred gene network using Espresso, a well-known logic simplification tool available from the University of California, Berkeley [19]. Through our automated tool, we generate a netlist for the minimized gene network. In our model, nodes are asynchronously updated. Boolean kinetics-based logic components (AND-, OR-, NOT-Gates) are dynamically implemented to accommodate variable numbers of gene inputs [24]. We use Xyce to simulate the eighty-one node (each node is a meta-gene) yeast gene network. Results were compared to the original discretized signal. Figure 7 shows the simulation output for meta-gene zero compared to the discretized microarray time series expression data.

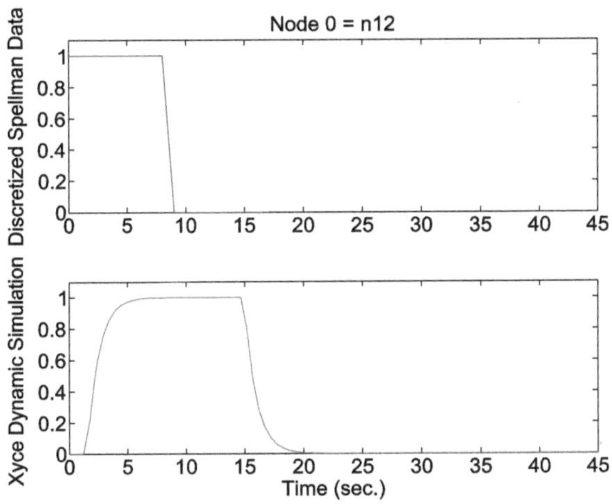

Fig. 7 Comparison of the output of Xyce simulation (top graph) to discretized time series microarray data (bottom graph) for meta-gene node zero

The two graphs are similar, which is expected given that meta-gene zero has a simple gene network. A noticeable difference is the width of the two curves, which can be addressed by adjusting the threshold point and parameters used in the Boolean kinetics model of the logic gates. Figure 8 shows the simulation output of meta-gene node forty-four and its corresponding input nodes and a comparison of the Xyce simulation to the discretized microarray data. For this more complex example, it is difficult to find similarities between the nodes (except for node zero, previously discussed). We suspect that some of the observed variations between the simulation and discretized data are due in part to the fact that the discretized data represents eighteen discrete time points whereas the simulation data represents continuous time steps. Application of a parameter estimation tool, like *Dakota*^{TM} [25], will help tune simulation parameters to produce results consistent with the discretized microarray data.

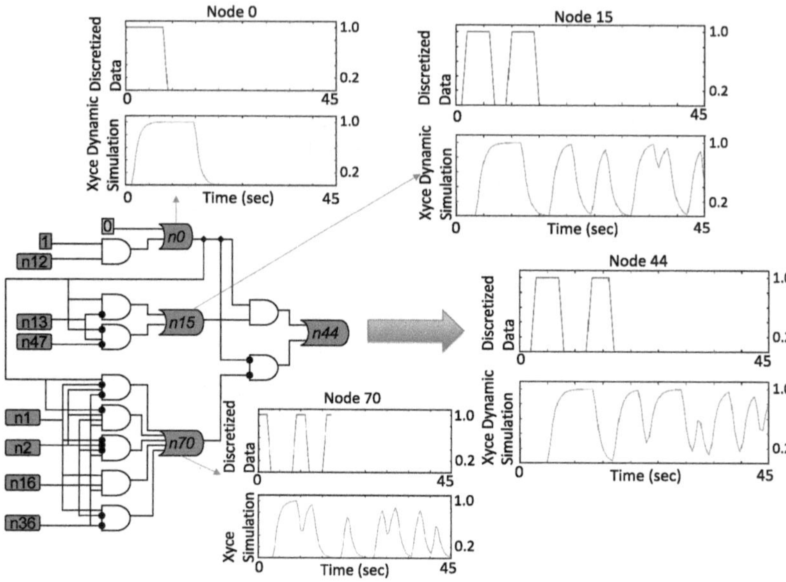

Fig. 8 Comparison of Xyce simulation results (top graphs) for meta-gene Node 44 and Nodes 0, 15, and 70 (inputs to Node 44) to discretized microarray data (bottom graphs)

4 Signal Transduction Cascades

Signal transduction networks enable the cell to sense and respond to its extracellular microenvironment. A cascades of signal transduction reactions can be initiated by the binding of a chemical signal to a cell membrane receptor, transduced through reaction pathways in the cytoplasm and result in the modulation of gene transcription and protein translation events. Many of the reaction steps in signal transduction can be represented using mass-action kinetics, as the concentration of the substrates may not be significantly larger than enzymes that may be involved in the reaction [26]. In this section, we demonstrate the use of BioXyce to model signal transduction cascades in the interferon gamma (IFNγ) based activation of the Janus kinase (JAK) and signal transducer and activator of transcription (STAT) pathway, an important pathway in host-mediated immune response to pathogenic infections. Our simulation model is based on the Yamada, et al. model of IFNγ activated JAK-STAT system [26], and several of the key reactions used in the BioXyce model are depicted in Fig. 9.

IFNγ can modulate immune responses through the activation of the JAK-STAT pathway. In brief, IFNγ binds to receptors on the cell membrane and JAK binds to IFNγ-receptor complex (IFNR) on the intracellular domain of the receptor. This initiates a cascade of reactions including the binding of STAT1 to phosphorylated IFNR-JAK homodimers, phosphorylation of STAT1, and translocation of phosphorylated STAT1 homodimers into the nucleus. This initiates transcription of IFNγ re-

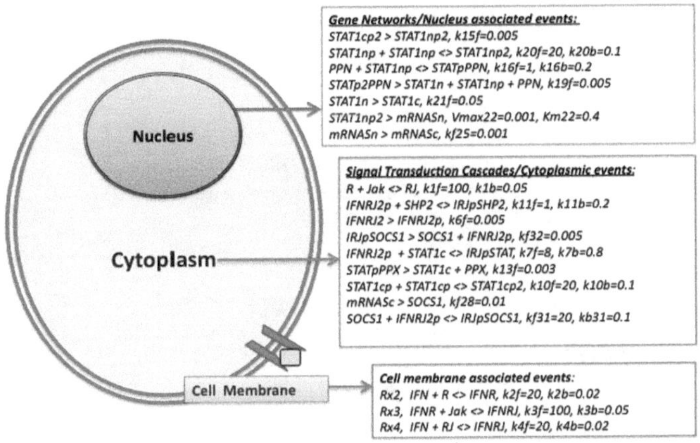

Fig. 9 Example reactions in the IFNγ activated JAK-STAT signal transduction pathway and relative location within the cell

sponsive genes as well as proteins, like suppressor of cytokine signaling-1 (SOCS1) that negatively regulates the JAK-STAT pathway. Cytoplasmic and nuclear phosphatases such as SHP-2 (SH2 domain-con- taining tyrosine phosphatase 2; cytoplasmic), PPX (cytoplasmic phosphatase), and PPN (nuclear phosphatase) also modulate the pathway by de-phosphorylating complexes and therefore regulating the signaling pathway. These are included in the BioXyce model with associated reaction kinetics as specified in the Yamada, et al. model [26] and are used to construct the signal transduction subcircuit and associated circuit netlist as shown in the excerpt below:

```
************** Signal Transduction Circuit **************************

xRxn_Rx2 IFN R IFNR stRxn2To1
    PARAMS: r1Stio=1 r2Stio=1 p1Stio=1 kRate={k2f}
** REVERSIBLE
xRxn_Rx2_Rev IFNR IFN R stRxn1To2
    PARAMS: r1Stio=1 p1Stio=1 p2Stio=1 kRate={k2b}

*********************************************************************
************** Mass-Action Reaction Subcircuit **********************

**** SigTrans Rxn1To2 ****
.SUBCKT stRxn1To2 rnt1 prd1 prd2
    PARAMS: r1Stio=1 p1Stio=1 p2Stio=1 kRate=1
* Modeling the simple mass-action reaction set of :
* r1Stio A1  -> p1Stio B1+ p2Stio B2
 Where* v = kRate* [A1] )
*
*** Reaction - consume the reactants
BReact1 rnt1 0 I={r1Stio*((kRate*((V(rnt1))**r1Stio)))}
*** Reaction - produce the products
BProd1 0 prd1 I={p1Stio*((kRate*((V(rnt1))**r1Stio)))}
```

```
BProd2 0 prd2 I={p2Stio*((kRate*((V(rnt1))**r1Stio)))}
.ENDS

**** SigTrans Rxn2To1 ****
.SUBCKT stRxn2To1 rnt1 rnt2 prd1
    PARAMS: r1Stio=1 r2Stio=1 p1Stio=1 kRate=1
* Modeling the simple mass-action reaction set of :
* r1Stio A1 + r2Stio A2 -> p1Stio B1
 Where* v = kRate* [A1][A2] )
*
*** Reaction - consume the reactants
BReact1 rnt1 0 I={r1Stio*((kRate*((V(rnt1))**r1Stio)
    *((V(rnt2))**r2Stio)))}
BReact2 rnt2 0 I={r2Stio*((kRate*((V(rnt1))**r1Stio)
    *((V(rnt2))**r2Stio)))}
*** Reaction - produce the products
BProd1 0 prd1 I={p1Stio*((kRate*((V(rnt1))**r1Stio)
    *((V(rnt2))**r2Stio)))}
.ENDS
```

The JAK-STAT netlist was executed using Xyce and Fig. 10 shows the results of the simulation. In Fig. 10 the horizontal axis is simulation time (tSim) and the vertical axis is concentration of the substrates involved. As the activated receptor-JAK complex is used, we see an increase in the cytoplasmic STAT1 phosphorylated homo-dimers, which translocates to the nucleus as evidenced by the increased level of nuclear STAT1 in Fig. 10 (first row, third graph). The second row of graphs show nuclear mRNA transcription, mRNA translocation to cytoplasm and translation into SOCS1. Using the BioXyce model and simulation platform we were able to capture the expected behavior of this pathway.

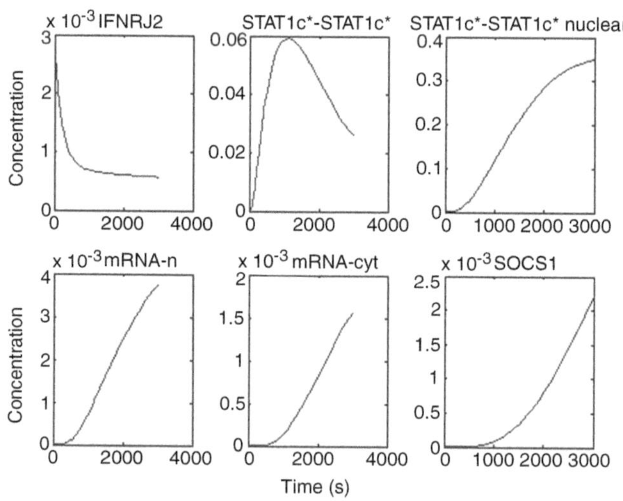

Fig. 10 Simulation results for the IFNγ modulation of the JAK-STAT signal transduction pathway

In this chapter we have briefly reviewed the development of BioXyce, a circuit-based simulation platform for systems biology. We have also demonstrated how to develop and represent models of biological networks using BioXyce and simulation using the Xyce electronic simulator engine. The BioXyce simulation platform has the demonstrated potential to handle the simulation of large and complex biological networks [2] and provides a computationally tractable approach for the *in silico* analysis of the macro-scale impact of micro-scale events on biological systems.

Acknowledgements The author is partially supported by Award Number K25HL075105 from the National Institutes of Health National Heart, Lung, and Blood Institute. The content is solely the responsibility of the author and does not necessarily represent the official views of the National Heart, Lung, and Blood Institute or the National Institutes of Health.

This project was funded by Sandia National Laboratories' Laboratory Directed Research and Development program. Sandia is a multiprogram laboratory operated by Sandia Corporation, a Lockheed Martin Company for the United States Department of Energy's National Nuclear Security Administration under contract DE-AC04-94AL85000.

References

1. S.R. Bolsover, J.S. Hyams, E.A. Shephard, H.A. White, C.G. Wiedemann, *Cell Biology – A Short Course, 2nd ed.* (Wiley-Liss)
2. E. May, R. Schiek, IET Systems Biology **3**(2)
3. L. Goldstein, E. Rypins, Comput. Methods Programs Biomed. **3**, 162 (1989)
4. H.H. McAdams, A.P. Arkin, Current Biology **10**, R318 (2000)
5. C.H. Schilling, J.S. Edwards, D. Letscher, Biotechnology and Bioengineering **71**(4), 286 (2000/2001)
6. M.W. Covert, C.H. Schilling, B. Palsson, Journal of Theoretical Biology **213**, 73 (2001)
7. S. Hoops, S. Sahle, R. Gauges, C. Lee, J. Pahle, N. Simus, M. Singhal, L. Xu, P. Mendes, U. Kummer, Bioinformatics **22**, 3067 (2006)
8. L.M. Loew, J.C. Schaff, Trends Biotechnology **19**, 401 (2001)
9. S.J. Plimpton, A. Slepoy, Journal of Physics: Conference Series **16**, 305 (2005)
10. H. Salis, V. Sotiropoulos, Y. Kaznessis, BMC Bioinformatics **7**(93) (2006)
11. M. Tomita, K. Hashimoto, K. Takahashi, T. Shimizu, Y. Matsuzaki, F. Miyoshi, K. Saito, S. Tanida, K. Yugi, J. Venter, C. Hutchison III, Bioinformatics **15**, 72 (1999)
12. C.R. Yang, B.E. Shapiro, E.D. Mjolsness, G.W. Hatfield, Bioinformatics **21**, 774 (2005)
13. Z.L. Xiu, A.P. Zeng, W.D. Deckwer, Journal of Biotechnology **58**, 125 (1997)
14. N. Yildirim, M.C. Mackey, Biophysical Journal **84**, 2841 (2003)
15. P. Smolen, D.A. Baxter, J.H. Byrne, Bulletin of Mathematical Biology **62**, 247 (2000)
16. A. Wagner, D. Fell, Proc. Roy. Soc. London Series B pp. 1803–1810 (2001)
17. D. Thieffry, R. Thomas, in *Pacific Symp. Biocomp.* (1998), pp. 77–88
18. H.H. McAdams, A. Arkin, Annual Review of Biophysics and Biomolecular Structure **27**, 199 (1998)
19. R.H. Katz, *Contemporary Logic Design* (1991)
20. A. Perelson, G. Weisbuch, Review of Modern Physics **69**, 1219 (1997)
21. P. Dhaeseleer, X. Wen, S. Fuhrman, R. Somogyi, in *Pacific Symp. Biocomp.* (1999), pp. 41–52
22. S.S. Shen-Orr, R. Milo, S. Mangan, U. Alon, Nature Genetics **31**, 64 (2002)
23. S. Martin, G. Davidson, E. May, M. Werner-Washburne, J.L. Faulon, in *ICSB* (2004)
24. A.S. Sedra, K.C. Smith, *Microelectronic Circuits* (1994)

25. M. Eldred, H. Agarwal, V. Perez, S.F. Wojtkiewicz Jr., J. Renaud, Structure and Infrastructure Engineering: Maintenance, Management, Life-Cycle Design and Performance **3**(3), 199 (2004)
26. S. Yamada, S. Shionoa, A. Joob, A. Yoshimurab, FEBS Letters pp. 190–196 (2003)

Analog Verification

Ken Kundert and Henry Chang

Abstract Just as digital design did 15 years ago, analog design has now reached a transition. The move to CMOS has made analog circuits more functionally complex, and that complexity leads naturally to functional errors in the designs, which in turn leads to re-spins and delays. And just as digital designers did 15 years ago, analog designers are beginning to realize that they need to employ a rigorous functional verification methodology. This chapter presents the basic concepts of analog verification and shows how it can be used to find a wide variety of functional errors in complex mixed-signal integrated circuits.

1 Analog Verification

Currently, 90% of all SOCs contain analog circuitry, and the analog content of these SOCs averages a relatively constant 20% of the area of the SOC. This analog content is implemented in CMOS and is relatively complicated, with hundreds and often thousands of digital control signals. Without a methodical and well designed verification process, the complexity of these circuits is resulting in increasing numbers of functional errors. Generally one or more 'test chips' are planned during the development of an SOC to test new circuits and architectures, tune yield, and to catch errors. However, unlike missing a performance goal, functional errors are problematic as they considerably reduce the value of the test chip. Functional errors degrade your ability to test and verify important aspects of your chip and software, and could make the test chip worthless as a demonstrator for your customers. It is these functional errors that are the primary focus of analog verification.

Today, very few design groups employ a systematic analog verification methodology [1, 3]. We are working to establish just such a methodology; one that has been

Ken Kundert and Henry Chang
Designer's Guide Consulting, Los Altos, California, USA
e-mail: ken@designers-guide.com, henry@designers-guide.com

P. Li et al. (eds.), *Simulation and Verification of Electronic and Biological Systems*, DOI 10.1007/978-94-007-0149-6_8, © Springer Science+Business Media B.V. 2011 well tested and shown to be both effective and practical [2]. It is this methodology that is described in this chapter. Concerning this methodology, if you are in charge of producing a chip, you might ask about the benefits and the costs. If you are involved in its design, you might ask about how it would affect you; will it be a burden or a help. If you are interested in becoming involved in the verification itself, you might ask if this is something that fits your skills and interests. This chapter attempts to give you the information to answer these questions at a conceptual level while filling in some of the details by way of a simple example.

Analog verification tends to find three types of functional errors. First, it finds errors within individual analog blocks that implement many settings. These errors are often subtle problems in the control logic. They are not found by the designer in cases where there are simply too many settings to test. For example, when designing a programmable gain amplifier with 64 gain settings, the typical designer will only simulate a few representative settings; the highest, the lowest, and maybe one or two in the middle. With this approach errors in the least significant bits in the gain control will likely go unnoticed. The second type of errors are inter-block communication errors; chicken and egg problems and the like. As an example, consider the case of a low-power on-chip regulator that is dependent on a shared bias generator. If the bias generator is itself dependent on the output of the regulator, then the pair may never start. These go undetected because blocks from different designers would have to be simulated together for them to be noticed, and that is rarely, if ever, done. The third type of errors is in the digital circuitry that controls the analog, produces its input, or processes its output; or it is in the interface between the analog and digital circuitry. They generally go undetected because it is usually very difficult if not impossible to co-simulate the analog and digital sections in any meaningful way (transistor simulations are much too slow). Analog verification addresses these potential errors with two new tools:

1. exhaustive self-checking testbenches based on Verilog-AMS [4], and
2. pin-accurate functional models written in either Verilog or Verilog-AMS.

To see the importance of both, consider the equalizer shown below in Fig. 1, as might be used as part of a high-speed digital data transmission system. During the course of the design the equalizer itself must be verified to function correctly. In addition, usually it is desirable to build a model of the equalizer that can be used when verifying the overall system. Analog verification fulfills both of these needs.

Writing an exhaustive equalizer testbench for a SPICE simulator would be very difficult, but Verilog-AMS provides a rich language that can be used to easily describe the needed tests. An example is given below. It consists of 20–30 lines of code that are tailored to the device under test and are used to thoroughly exercise it and confirm that it produces the expected output (for brevity, boilerplate code is not shown). The testbench operates by applying a unit impulse to the input, and then monitoring the output as the impulse propagates through the delay line. On the first clock cycle the output should produce k_0, on the second k_1, etc. After the unit impulse exits the delay line, the coefficients are changed and a new impulse is fed in. Notice that each coefficient is stepped through all of its 16 possible values, and each coefficient is always set differently from the others. This minimizes the likelihood that a wiring or logic error will be missed.

```
module testbench ();
reg in, clk = 0, vdd = 1, gnd = 0;
reg [3:0] k0, k1, k2, k3, k4;
electrical out;
logic2p5 in, clk = 0;
supply2p5 vdd, gnd;
integer i;

// Instantiate the device under test (DUT)
equalizer DUT(
    .out(out), .in(in), .clk(clk),
    .k0(k0), .k1(k1), .k2(k2), .k3(k3), .k4(k4),
    .vdd(vdd), .gnd(gnd)
);

// Generate the clock
always #1 clk = ~clk;

// Test the DUT
initial begin
    k0 = 0;
    k1 = 0;
    k2 = 0;
    k3 = 0;
    k4 = 0;
    in = 0;

    // clear out the delay line
    for (i=0; i<5; i=i+1)
        @(clk);
```

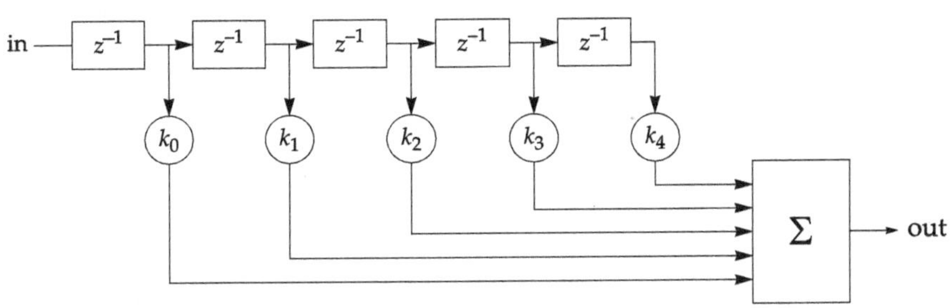

Fig. 1 An equalizer used in digital data communication

```
        // send an impulse through delay line
        // to check coefficients
        for (i=0; i<16; i=i+1) begin
            k0 = (i+0);
            k1 = (i+1);
            k2 = (i+2);
            k3 = (i+3);
            k4 = (i+4);
            in = 1;
            @(clk);
            in = 0;
            checkOutput(k0/(5.0*15), V(out), "k0", k0, 10m);
            @(clk);
            checkOutput(k1/(5.0*15), V(out), "k1", k1, 10m);
            @(clk);
            checkOutput(k2/(5.0*15), V(out), "k2", k2, 10m);
            @(clk);
            checkOutput(k3/(5.0*15), V(out), "k3", k3, 10m);
            @(clk);
            checkOutput(k4/(5.0*15), V(out), "k4", k4, 10m);
        end
        reportTestResults;
        $finish;
    end
            ...
endmodule
```

When run on the circuit as shown in Fig. 2, the testbench produces the waveforms shown below in Fig. 3. However, one does not need to actually view the waveforms to determine if the circuit is working correctly. This testbench runs 80 separate tests on the equalizer, all while driving each of the inputs independently through all possible values. When run on the circuit, it will produce a summary output that indicates which tests the circuit passed, and which it fails. If the testbench is written exclusively from the functional specifications for the block, if the tests are comprehensive, and if they all pass, then the circuit implementation has been verified to be functionally equivalent to the specification. This in itself is generally much more verification than is done for analog blocks today. However, this is only the first of three sets of tests that are run, verifying that each individual analog or mixed-signal

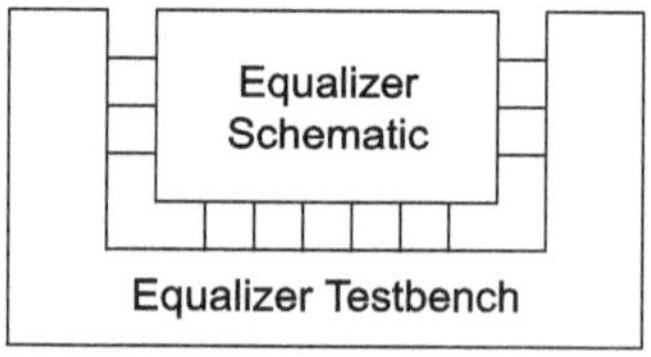

Fig. 2 Testbench applied to schematic

block is implemented correctly. The next two sets will verify that the blocks operate together as expected and that the entire analog subsystem inter-operates properly with the digital subsystem.

To take the next step, it is necessary to have a pin accurate functional model of each analog or mixed-signal circuit block in the analog subsystem. The model for our equalizer is shown below.

```
module equalizer(
    out, in, clk,
    k0, k1, k2, k3, k4,
    vdd, gnd
);
output out; electrical out;
input in, clk, vdd, clk;
input [3:0] k0, k1, k2, k3, k4;
reg z0, z1, z2, z3, z4;
logic2p5 in, clk;
supply2p5 vdd, gnd;

// delay line
always @(clk) begin
    z0 <= in;
    z1 <= z0;
    z2 <= z1;
    z3 <= z2;
    z4 <= z3;
end

// weighted summer
integer result = 0;
reg fault = 1'b0;
always @(*) begin
    result = z0*k0 + z1*k1 + z2*k2 + z3*k3 + z4*k4;
    fault = ^result === 'bx || vdd !== 1 || gnd !== 0;
```

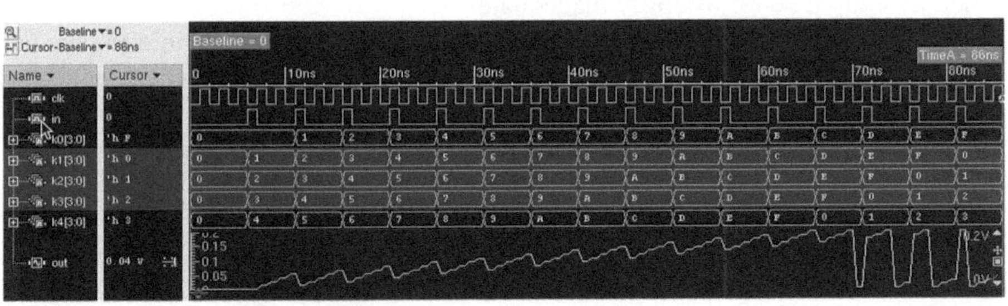

Fig. 3 Response of equalizer to tests

```
    if (fault) begin
        // set output to 0 if there is a fault
        result = 0;
    end
end
analog V(out) <+ transition(result/(5.0*15), 0, 50p);
endmodule
```

While very simple, this is a complete functional model of the equalizer that can be used in a full mixed-signal simulation. Notice that all of the pins are included in this model, including `Vdd` and `gnd`. Furthermore, notice that their effect on the output is modeled, albeit in a very crude manner. The model generates a zero output if power and ground are not properly supplied. The same approach would be taken for bias and reference inputs if they were present. This is important as it assures that the regression tests will fail if the signals the block requires are not properly provided. As such, this model is referred to as being *pin-accurate*.

Once the model is written, there is the question "does it faithfully represent the circuit?" At this point the question is easy to answer with authority. Simply run the testbench on the model, as shown in Fig. 4. Since the model is pin accurate, it is a simple matter of changing the configuration of the simulator to swap out the implementation view of the block for the model view. If all tests pass, the model and the circuit are functionally equivalent.

It is important to recognize that while it was easy for us to confirm that the model matched the circuit, it was only possible because we invested in building a comprehensive testbench. Most people that write models for their blocks do not perform this verification because they did not develop a testbench. This is a very dangerous situation as errors in the model may not be found and those errors may result in errors being injected into the implementation of surrounding blocks to compensate for the errors in the model.

In practice, the model and testbench are developed together. When the they are complete and all the tests pass, the model view of the block is replaced with the implementation view. This seemingly minor clarification is actually very important. The model simulates much faster than the implementation, making it much easier to add new tests to the testbench and check that they work as expected, which results in a testbench that is much more complete and refined. Without models, testbenches generally end up being woefully inadequate.

With a verified model of the mixed-signal blocks, it is now possible to take the next step of verifying that all of the analog blocks operate properly when connected

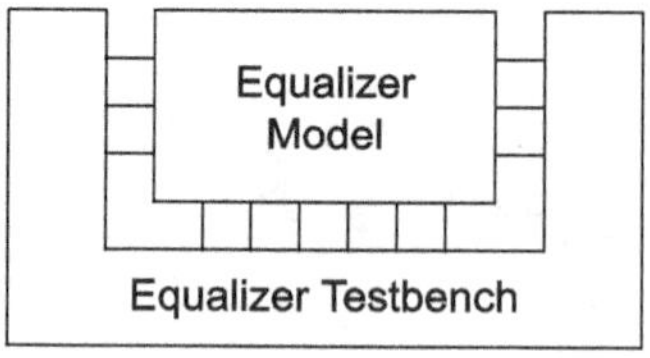

Fig. 4 Testbench applied to model

together. To do so, one writes a testbench for the entire analog subsystem and applies it to the top-level schematic of the analog section, where each block is represented by its fully verified model. For our example, one might combine the equalizer with the line driver and the receiver and perform loopback testing as shown in Fig. 5. This would be much too expensive if all of the blocks were at the transistor level, but is very reasonable when all of the blocks are represented by functional models. For more complete testing, one can drop one block at a time to the transistor level while continuing to use models for the remaining blocks. This is referred to as mixed-level simulation. While more expensive than model-level simulation, it does provide additional benefits.

With the models already created it is often possible to verify that the analog and digital subsystems operate together as expected. Simply write a testbench for the entire system and simulate both the analog, represented at the model level, and the digital, represented with RTL, together in a Verilog-AMS simulator. While this is the simplest solution, it can be problematic in certain cases. It may be that the analog models are too slow to allow a thorough top-level verification, or it may be that the there are constraints on the top-level verification, such as the need for System-Verilog, that cannot be satisfied with existing Verilog-AMS simulators. In these cases a new model is created, often a model written purely in Verilog. Again, this is a functional model, and so is generally not difficult to write. And again, the existence of a testbench means that the model can be verified to match the implementation, as shown in Fig. 6. In this case, the testbench is likely incompatible with the model as analog interface signals cannot be modeled directly in Verilog. There are several ways to model analog signals within the limitations of Verilog including pulse-with modulating a digital signal or using hierarchical references. To overcome the incompatiblity between the Verilog-AMS testbench and the Verilog model, the Verilog model is wrapped with a collar module whose sole purpose is to adapt the true analog signals from the testbench to something that can be accepted by Verilog. This collar is removed when the model is used for full-chip verification.

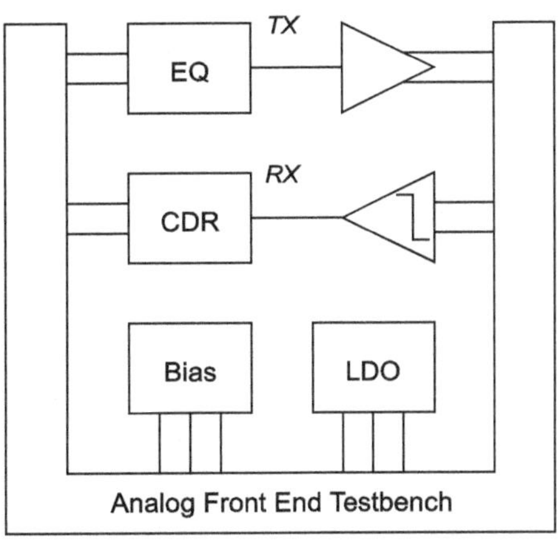

Fig. 5 Testbench applied to entire analog front end

The importance of verifying the model cannot be overstated. As mentioned before, without verification the model could contain errors, which creates the risk that an otherwise working design will be modified to properly operate with the model, thereby breaking the implementation.

Finally, the verified Verilog model of the analog front end is used with the chip-level testbench to verify that the analog and digital sections work together as expected, as shown in Fig. 7. An interesting question at this point is who should perform the top-level verification: the analog verification engineer, the digital verification engineer, the designer engineers, of the system engineer. What we have found is that it is best if all are involved. Typically the analog and digital verification engineers cooperate to build the basic testbench and tests. Then the design and system engineers use this testbench as a starting point for more targeted testing. They modify the testbench and tests to more completely exercise their portion of the design to assure that it is driven and responds as expected within the context of the whole chip. After all, it is they that are the most able to recognize subtle issues.

Using the "top-down" and "bottom-up" terms, we describe this methodology as architecturally and functionally top-down with performance designed in a bottom-up style. In general, design teams already follow this approach. The difference is in

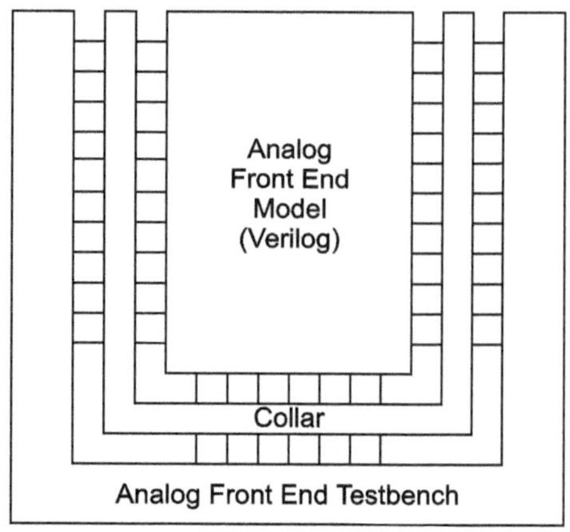

Fig. 6 Testbench applied to Verilog model of the analog front end. The collar adapts the existing testbench to the Verilog model

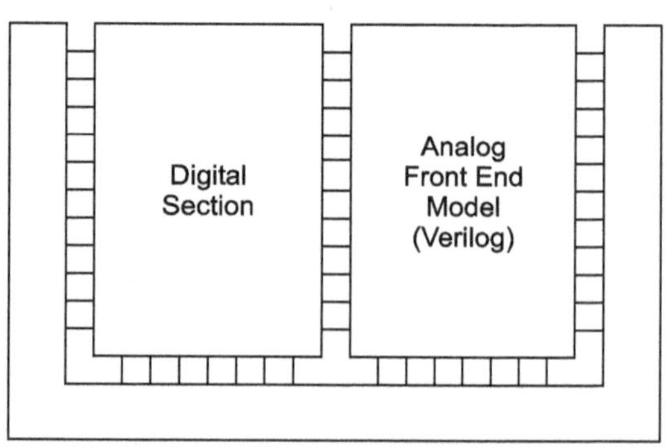

Fig. 7 Testing the entire chip

the level of solidity in the functionality going down. We claim that with just a bit more definition, huge benefits can be derived with verification.

2 Design Time Line

The design project begins with the creation of the specifications for the analog portion of the design. These are generally derived by the chip system engineer. Once the chip architecture is solidified, the system engineer works with the analog design lead to formulate specifications for the analog portion of the design based on an estimate of what is possible in the analog section. Reuse of existing components is often taken into account.

Analog architectural questions are explored such as determining the best architecture for blocks and what are the detailed parametric design values to be used. System exploration tools such as Matlab, Simulink, Ptolomy, and even a generic spreadsheet are helpful in this phase because they allow designers to quickly explore their design and to transform the design requirements into design parameters. For example, Matlab is adept at converting filter characteristics into filter coefficients. At this point in the design flow the focus is on modeling the signal flow through the system with an emphasis on estimating expected performance, so second-order effects are often included. However, details such as modeling the control flow or interface details between the blocks are rarely modeled, because it would slow down architectural exploration.

This is the time when the analog verification lead joins the project. He/she works with the design lead to develop models for each block in Verilog-AMS. Verilog-AMS is used rather than a language like Matlab because these models will eventually be used to perform mixed-level simulation. Here the modeling focus changes from the system design phase to the implementation phase where verification is a key goal. Only the function of the blocks is modeled with little to no effort expended to modeling second-order effects. Instead the emphasis is on modeling all of the functionality, including control flow and the interfaces. Thus, system exploration tools are the tool of the design engineer, while AHDLs are for the verification engineer. At this point, the verification engineer can also influence the design to make it more verifiable and testable. The outcome of this phase is a first cut at a partitioned pin-accurate top-level analog schematic with behavioral models for each of the blocks that is simulatable and implements all functional aspects of the specification. The block designers now join the project.

The block designers start with the documentation, the top-level schematic, and the models, and use them to gain an understanding of what they are expected to design. Meanwhile, the verification engineer or engineers (more would join the project at this point if it is large or complex) would begin developing block-level and then analog system-level tests. This is where rigor is brought into design management. Two types of tests are developed. Quick tests are a set of basic tests that can ideally run in minutes and can be used by designers before they "check-in" their blocks to assure that they meet minimum functionality and compatibility requirements. Regression tests are more extensive, designed for overnight runs. It is important for designers to check in their designs often so that the verification engineer can run these more extensive tests. It also helps to keep the overall design synchronized. With the designers' focus on performance, they will likely also use the models written by the verification engineer to speed up simulation by replacing noncritical circuitry with simple functional models. For higher level specifications such as bit error rate (BER) and signal-to-noise ratio (SNR), more complex verification post-processing can be included. The post processing can be custom written or the system exploration tools can once again be used.

Besides developing and running tests, the verification engineers also develop and test the Verilog model for the full analog section if one needs to be delivered to the customer. They would also be expected to perform verification reviews that include themselves, the block designers, the analog lead, and CAD support to assure that their tests are both comprehensive and efficient.

Once the top-level schematic and models have settled out it is possible to bring the test engineers onto the project. They would use the models to begin developing the tests that will be used on the production floor. Bringing the test engineers in early allows the tests to be developed in parallel with the design effort. This can shorten first customer shipments and allows them to contribute suggestions that would make the design more testable.

As the design approaches completion, the focus of the verification engineer shifts to more comprehensive system-level tests. Every block should be simulated at the transistor level within the larger system in a mixed-level simulation. If there is a possibility of subtle interaction between multiple blocks, the blocks should be simulated together at the transistor level. An example of when this is necessary is when the design includes a separate bias block. That block should be run pair-wise at the transistor level with every block it feeds.

Maximum benefit is derived from the verification tasks when verification starts at the beginning of the project because the design benefits from automated regression testing throughout most of the project.

2.1 Incomplete Implementation of the Methodology

The verification process is affected if at the start of the project, the design lead cannot (or chooses not to) define the architecture, functionality, and the interfaces in enough detail that the top-level models and testbenches can be developed. The level of effort is virtually identical. However, the difference in the amount of verification that can occur is dramatic. In the ideal case, during the entire design phase there is benefit from verification. In the late decision case, benefit is only derived during final layout. There are several disadvantages. Design is slower because designers need to take the time to continually check functionality. Also, errors that are caught by verification result in a costly engineering change order (ECO) where layout has to be changed.

For maximum benefit, the design lead must be willing to make an early decision on interfaces and block functionality. This does not mean these decisions are set in stone. As long as these interfaces do not change on a daily level, the verification methodology will work. Making early decisions is usually to the benefit of the lead designer. Too often, if these decisions are not made, the block designers end up making the interface decisions and the lead designer is constantly behind trying to keep the design synchronized while the block designers continue to make changes. The lead designer is also the one that derives most of the benefit from this methodology—as the person who is ultimately responsible for ensuring that the analog section is designed correctly.

3 Analog Verification Engineers

Analog verification requires a change in the way things are done in most analog design groups. New skills must be learned and new types of engineers must be found, hired, and trained. Why would one go to all this trouble? The answer is easy: there is no other way to assure your design will function properly before you build it. Despite what simulation vendors may want you to believe, exhaustive regression testing on transistor level schematics is completely impractical. Furthermore, it will never again in the foreseeable future be practical because the complexity of the circuitry is increasing faster than the speeds of the simulators and computers they run on. It is important to realize that the complexity of analog circuits increases in three independent ways simultaneously: the circuits become larger, they become algorithmically more complex, and the number of modes and settings they support increases; all of which were accelerated by the switch to CMOS.

Once adopted, our approach to analog verification is seen as being preferable to the old ways that only involved transistor simulation in two ways. First, it is based on the use of models, and so can occur much earlier in the design cycle. It can find errors before the transistor-level circuits are designed, which can save significant design effort. Second, the regression tests themselves are of much higher quality when they are developed in concert with the models. Because the models run so much faster, they allow the tests themselves to be much more fully exercised and tested. Regression tests developed in concert with models are more comprehensive and more sophisticated that those developed in concert with transistor-level circuits alone.

To undertake this methodology one needs engineers trained in the art of analog functional modeling and testing and focused on verification. The alternative, using designers to both design and verify, is generally not as successful. It can be difficult to convince them to do it, they generally do not have the skills, and they tend to prioritize design activities over verification activities.

We have found that once an analog verification methodology is established, it requires around one verification engineer for every five design engineers (a number that varies depending on the total number of modes and settings implemented by the circuit). Successful analog verification engineers need a variety of skills. They must have an analog background, meaning that they understand and are comfortable with analog circuits. They must be modeling engineers in that they must be facile with analog and digital modeling languages, such as Verilog-AMS, and they must know how to write models and testbenches that are accurate, robust, and efficient. They must be hackers in that they can quickly write scripts and develop workarounds to issues with the design tools. And finally, they must be verification engineers, meaning that they must be detail oriented as well as natural skeptics.

Analog verification engineers are currently very rare, and so must be developed. To that end, we provide both training and guidance as a way of helping to establish the discipline. Fortunately, being an analog verification engineer is a desirable occupation. It is a creative endeavor that allows one to be involved in the design of analog integrated circuits, but provides more variety than if one were to design the individual blocks that make up the chip.

4 Adoption

Analog verification is a large change for design groups, and adopting it represents a substantial investment. Success in adopting the methodology is assured only if the design company fully commits to it. As a result, it is almost always necessary that the change be driven from the top by the business owner; the one that both has the responsibility to deliver a working chip and authority to dedicate the needed resources. It is the desire to control the risk of a functional failure in the chip that drives adoption. However, the motivation to control this risk often only comes after a spectacular failure.

When first adopting an analog verification methodology, it is generally best to allow the verification effort to trail the design effort. The focus during this initial phase is on developing both expertise and a cache of models and testbenches that can be used on subsequent designs. It is during this phase that designers begin to first trust and later count on the verification engineers. It is also when the extent of the testing possibilities starts to become clear. Designer's have been so constrained in what they can test for so long, that it takes a while before the magnitude of the change really sinks in.

After the design and verification teams become comfortable with the analog verification methodology, and once a cache of models and testbenches has been established, then it becomes possible for verification to actually lead design. The models, testbenches, and top-level schematics are developed before the individual blocks are designed. In this way, verification can actually improve the efficiency of your design team. It does so by flushing out many of the errors and ambiguities in the specification and architecture before the expensive process of transistor-level design begins. In this way, analog verification naturally leads to an increased use of top-down design.

Another interesting side effect of adopting analog verification is that it tends to lead to healthier design teams. Analog verification provides a safety net that catches the errors of inexperienced design engineers making it possible to build design teams from designers that have a range of experience levels. No longer is it necessary to limit yourselves to only hiring design engineers with five years experience, which greatly increases the pool of available engineers and allows engineers to develop naturally in your company, increasing retention.

5 Examples

To give you a sense of what is possible, several representative verification efforts are described.

5.1 Audio Codec

As part of a larger SOC, the stereo audio codec is responsible for converting the input signals from a pair of microphones into a data stream and converting a data stream into an output signal that drove a pair of speakers or headphones. It consisted of a pair of digitally controlled pre-amplifiers and a pair of high resolution $\Sigma\Delta$ analog-to-digital converters on the input side and a pair of $\Sigma\Delta$ digital-to-analog converters and a pair of power amplifiers on the output side. Also included were power and bias supplies and a digital serial interface to the rest of the SOC. Although a relatively simple design, it consisted of over 25 analog blocks for which models and testbenches were needed.

In this example the verification effort was started late and the design reached tapeout before the verification methodology started producing results. As such, we can conclusively say the errors found during functional verification would have made it into the test chip. Two types of errors were found. First, errors were found within a single block; in this case errors in the control logic. These errors were not found during the design phase because the errors only affected the least-significant bits and the large number of settings precluded exhaustive testing of every setting. Second, errors occurred because of mutual dependencies between two or more blocks. In one case, the sign of the bias current was reversed. In another, two blocks depended on the output of the other. In both cases the errors would have resulted in the those blocks failing to operate.

With all models in place, the top-level schematic of the audio codec was simulated. Each setting required only a minute or so to check, even though the loop contained several $\Sigma\Delta$ converters. This effort represented the first time that this company had ever been able to perform a full top-level simulation of this type of design.

5.2 Micro-Controller Based Power Management Unit

This design includes a half dozen switching regulators and a dozen linear regulators. It also contained a micro controller. Communication to the controller occurred through a serial IO bus. The verification effort found errors in the analog blocks, in the RTL, and in the software. Careful modeling allowed the simulation of a full boot sequence plus several seconds of real time operation in only 30 minutes. Such a simulation allowed the design lead to observe the full operation of the chip before it was built, and resulted in changes to the software to optimize its start-up behavior.

5.3 RF Transceiver

RF transceivers are particularly difficult to verify because of the wide range of frequencies present in the design. Again, careful modeling allowed a full loopback simulation of the transceiver that included the RTL and the TX, RX and the synthesizer for multiple packets. Besides finding errors in the analog blocks and in the analog/digital interface, errors were also found in the RTL. The presence of the RF signal path allowed the RTL to be exercised and observed in a way that was not possible when the RF path was missing, and this exposed several serious issues in the RTL itself.

5.4 SerDes

The serializer-deserializer (SerDes) of a high-speed serial interface is basically a digital circuit, and so it would seem that digital verification would be used to assure it functioned as expected. However, the very high speeds that these circuits operate at (40 Gb/s for OC-768 optical networks) prevents the use of standard-cell libraries, making them incompatible with traditional digital verification. However, analog verification can be applied to these custom digital designs. When we did so, we found simple logic errors, a reversed serial-to-parallel conversion, and errors in the RTL of the low-speed control logic.

6 Conclusion

Just as digital design did 15 years ago, analog design has now reached a transition. The move to CMOS has made analog circuits more functionally complex, and that complexity leads naturally to the need for analog functional verification. It is inevitable that all mixed-signal system-on-chip design groups are destined to adopt analog verification. That being said, at this point it is not an easy transition. There are many barriers to be overcome and pitfalls to be avoided. It is easy to get frustrated and give up. However, know that it does work, and those that successfully adopt analog verification quickly reach the point where they cannot imagine designing without it. There are two factors that if present go a long way towards ensuring success in adopting analog verification: experience and commitment. We help companies transition to analog verification by providing experience in the form of training and guidance. The commitment is up to you.

You can access a fully complete and executable version of this circuit and testbench at `www.designers-guide.com/newsletter/0807/example.tgz`.

References

1. Henry Chang and Ken Kundert. Verification of complex analog and RF IC designs. *The Proceedings of the IEEE*, February 2007. Available from `www.designers-guide.com`.
2. Ken Kundert. Analog verification. *IEEE Custom Integrated Circuits Conference*, Educational Session, September 2007.
3. Ken Kundert. Principles of top-down mixed-signal design. Available from `www.designers-guide.org`.
4. Ken Kundert and Olaf Zinke. *The Designer's Guide to Verilog-AMS*. Springer, 2005.

Formal Methods for Verification of Analog Circuits

Sebastian Steinhorst and Lars Hedrich

Abstract This chapter will discuss how the demand for higher design and verification efficiency could be satisfied by introducing formal methods into analog circuit verification. Motivating the need for formal methods by a practical design example where conventional verification failed to detect a critical circuit property, an overview over the history and research in progress on different formalized analog verification approaches in the areas of analog model checking, equivalence checking, and formalizing test bench simulation approaches is given. Furthermore, a framework for unifying formal verification of analog circuits is presented and applied to example circuits. The proposed framework is based on the Analog Specification Language (ASL), the corresponding verification algorithms, and an approach to generate complete state space-covering input stimuli for transient simulation with guaranteed coverage of the complete reachable state space of the circuit under verification.

1 Introduction

The impact of electronic devices on our everyday life is inevitable. With the dependency on electronics increasing continuously, the consequences of errors in such electronic systems are increasing just as well. There are several levels of severity from just being disconnected during a phone call to possible airplane crashes due to errors in the electronic components. Even for non-safety-critical cases, errors in

---

Sebastian Steinhorst and Lars Hedrich
Electronic Design Methodology
Institute for Computer Science
Goethe University of Frankfurt/Main
Robert-Mayer-Str. 11-15
D-60325 Frankfurt/Main, Germany
e-mail: [steinhorst,hedrich]@em.cs.uni-frankfurt.de

P. Li et al. (eds.), *Simulation and Verification of Electronic and Biological Systems*,
DOI 10.1007/978-94-007-0149-6_9, © Springer Science+Business Media B.V. 2011

electronic systems have an economic dimension, where the cost of missed design flaws is determining whether a company can stay competitive or not.

Due to the increasing system complexity and decreasing time to market, verification of analog circuit designs has become a more and more crucial part of the electronic circuit design flow. While formal verification methods are established in the digital domain, industrial analog circuit design flows are lacking formal or at least formalized verification methodologies. Analog circuit verification still depends on the designer's experience and expertise to manually define appropriate test benches for simulation-based design flows and to select the right input signals in order to detect possible design errors. Driven by the perennial demand for higher design efficiency, new approaches offering more automation of the verification process are of vital importance.

There has been significant progress in several areas of electronic design automation (EDA) for analog circuits. Some complex tasks such as placement, sizing, and design centering have been addressed by EDA-vendors, now being available as automated tools which are fully integrated into the design flow. These tools are exploiting algorithmic concepts which by far outperform manual approaches.

By contrast, the area of analog design verification is not systematically covered by existing tools yet. While approaches to assertion-based simulation are emerging, which are mainly automating previously manual efforts, they are not targeting the fundamental problem of analog circuit verification: verification coverage. Today's established common verification methodology is analyzing the circuit's behavior by simulation using test benches. Specification conformance is checked by performing several transient simulations with input signals which are considered representative for the future operating conditions of the circuit. Although this approach to discover design errors has been working for decades, redesigns have occurred frequently due to missing some critical behavior of the circuit during simulation.

The approaches to formal verification of analog circuits have the common characteristic of targeting the verification coverage problem of user-defined transient simulations. Figure 1 illustrates the problem of conventional transient simulations not covering the complete reachable state space of a circuit and summarizes the main characteristics of this approach. In contrast, formal verification approaches cover the complete state space of the circuit, resulting in absolute confidence concerning the verification results, but with the downside of requiring new verification paradigms.

2 The Need for Formal Methods

The following motivating example will illustrate that conventional analog circuit simulation within a test bench setup can lead to wrong verification assumptions and how, in contrast, a complete formal verification using some of the methodologies described in this chapter can identify hidden design errors.

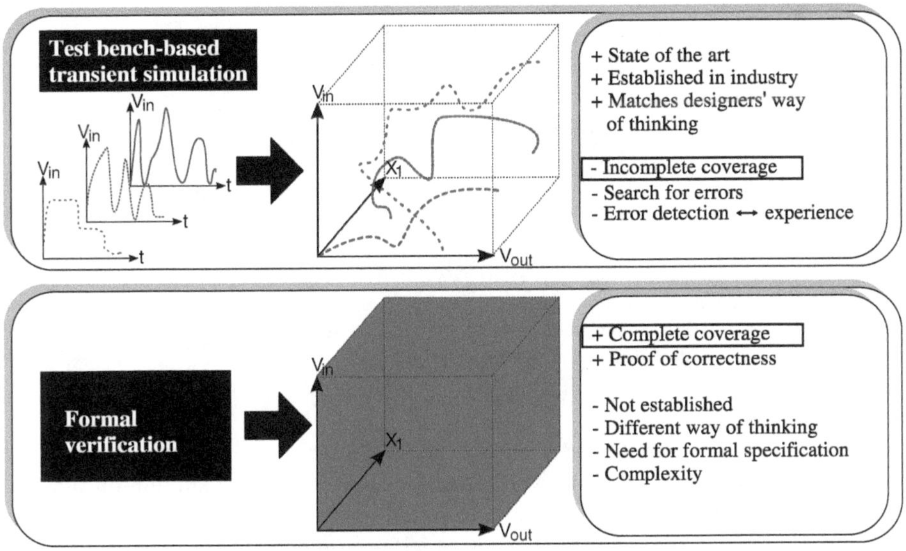

Fig. 1 Comparison of verification by test bench-based transient simulation and formal verification

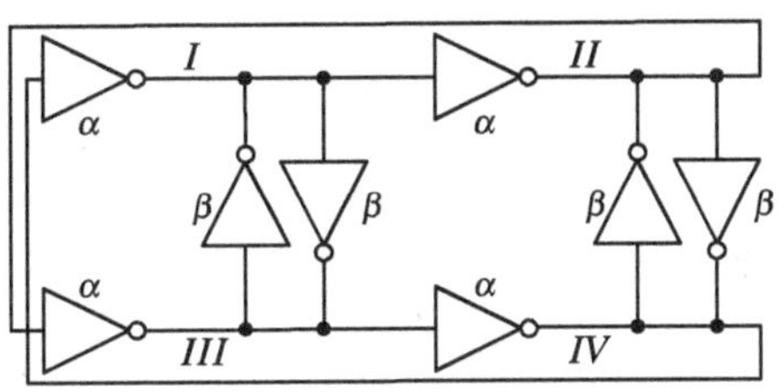

Fig. 2 Modified ring oscillator with an even number of inverter stages and cross-coupling

When a finite number of simulations with input stimuli or initial conditions is conducted and the expected behavior of the circuit is validated by the simulations, in today's industrial verification applications, the circuit is assumed to be successfully verified. However, this is only due to the lack of formal verification tools. As pointed out in the previous section, not finding a specification violation with simulation runs cannot prove that the specification is satisfied under any circumstances.

The example circuit illustrated in Fig. 2 is a modified ring oscillator with an even number of inverter stages and cross-coupling [15]. Due to the bridges β, the circuit oscillates if there is a ratio α/β of the transistor sizes in the feedback chain to those of the bridges within the interval [0.4, 2.0].

This circuit has in fact been considered as successfully verified by transient simulation with a set of predefined initial conditions and went into production. What was discovered only after the tape-out of the circuit is its crucial property of be-

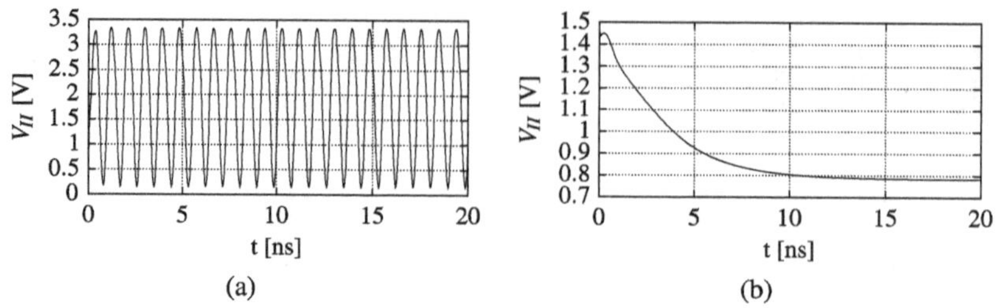

Fig. 3 Transient responses for the node voltage V_{II} of the ring oscillator for transistor ratio $\alpha/\beta = 1.95$. With initial conditions $V_I, V_{II}, V_{III} = 0$ V and $V_{IV} = 0.5$ V, the circuit oscillates (**a**). With the initial conditions $V_I = 3.33444$ V, $V_{II} = 1.49605$ V, $V_{III} = 3.16195$ V, $V_{IV} = 0.207917$ V detected by formal verification, the circuit does not oscillate (**b**)

ing prone to certain initial conditions that prevent it from oscillating when the α/β ratio reaches or exceeds the interval boundaries. While several simulation runs for this critical transistor ratios with random initial conditions can show perfect oscillation, particular initial conditions exist which are preventing this circuit from oscillating. Such simulation runs for two different initial conditions are illustrated in Fig. 3.

The critical behavior with certain initial conditions could have been detected using a formal property verification approach. In order to demonstrate this, a formal verification of the circuit using a property specification in the Analog Specification Language (ASL) and the corresponding implementation in a model checking tool [23] (which will be introduced in Sect. 4.2) was performed. Therewith, for transistor ratio 1.0, no initial conditions violating the oscillation behavior are reported by the verification algorithms.

In contrast to the perfect oscillation with transistor ratio 1.0, for transistor ratio 1.95, initial conditions are detected by the formal verification method for which the circuit will not run into an oscillation. This area of nontrivial bad initial conditions is illustrated in Fig. 4 projected to the state space variables V_I, V_{II}, and V_{III} and the non-oscillating behavior shown in Fig. 3b was identified by a transient simulation starting from the set of these initial conditions.

3 Overview over Formalized Analog Verification Methods

3.1 Analog Model Checking

The first approach to model checking of analog circuits introduced a discrete state space model of the circuit to be checked against a property specification given as ω-automata. The circuit investigated was a transistor level representation of a digital interlock, described using simplified transistor models implemented by a capacitor and a voltage-controlled current source [16]. Although the discretization into

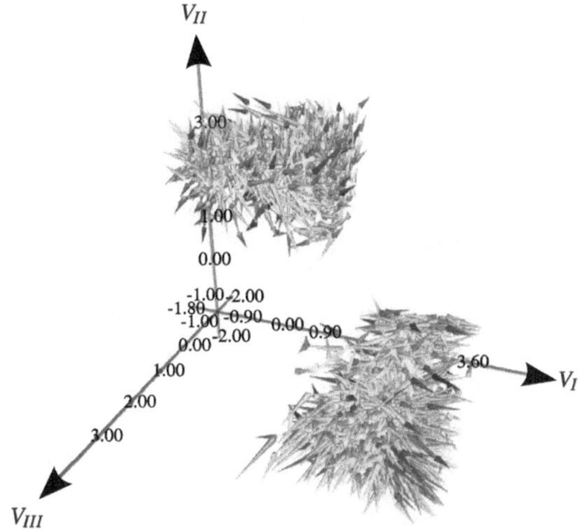

Fig. 4 State space trajectories of initial conditions leading into the non-oscillating steady states for transistor ratio $\alpha/\beta = 1.95$

a finite automaton is very rough, this is the first approach to discrete modeling of the state space of analog systems with application of an automata-based property verification.

Derived from the successful property specification of digital and software systems with the Computation Tree Logic (CTL) [3], the first approaches to CTL-based model checking of analog systems applied CTL extended with an analog operator to a discretized state space model of nonlinear analog circuits [13]. With an adaptive state space discretization that controls the size of the enclosures of state space regions depending on the level of homogeneity of the state space dynamics, basic properties of nonlinear circuits have been verified. The addition of time constraints to the CTL specification allowed for the first time to formally verify timed behavior of analog circuits [9]. In [5, 6] a verification system applying an extended CTL derivative called AnaCTL to the transient response of analog circuits is proposed, weakening the formality of the approach. By higher level modeling of analog circuits using labeled hybrid petri nets, verification of AMS-systems is introduced in [17, 18], not offering an automated translation of transistor level circuits to the petri net representation. The same limitation applies to the approach presented in [1], where a stability verification of a symbolic mathematical representation of a delta sigma modulator using recurrence equations is introduced.

Despite the different developments, the specification of analog properties and the corresponding verification algorithms still remained limited by temporal logics. To target this limitation, specification and verification of complex analog circuit properties using specifications in ASL for checking on discrete state space models was introduced in [23, 26].

3.2 Analog Equivalence Checking

A state space sampling-based approach for equivalence checking of nonlinear analog circuits was initially introduced in [11] with application to different CMOS inverters and extended in [14] for application to a Schmitt-trigger implemented as transistor netlist and behavioral description. For both system implementations under verification, using a local linear transformation for each sample point to a canonical representation (Kronecker's canonical form) and by numerically integrating these linear local transformations, an approximation of the nonlinear transformation for the system is obtained. The numerical differences between both internal transformed dynamics and the output variables provide a direct measure for the equality/difference of both systems.

Approaches for linear analog circuits with parameter tolerances were presented in [12] and [21].

Based on the PVS theorem prover, in [10] the functional equivalence of behavioral descriptions in VHDL-AMS and their synthesized analog circuits is checked by evaluating piecewise linear approximations of the analog behavior in the DC and low frequency domain. However, due to the simplifications, this approach is not a complete verification methodology.

Generation of complete state space-covering input stimuli enabling transient simulation with guaranteed coverage of the entire reachable state space of the circuits under verification is proposed in [22]. By automatically comparing the transient responses of two circuit implementations to such stimuli, equivalence checking is possible with guaranteed certainty on the results due to the complete coverage of the circuits' behavior [25]. Unlike other approaches to analog equivalence checking, the possible level of abstraction of behavioral models that can be compared to transistor netlists is not restricted.

3.3 Formalizing the Verification Flow

With the growing need for formalizing analog verification in industrial design flows and analog formal verification tools not yet being available, approaches that give up formality in favor of delivering practical solutions have emerged. Systematic verification based on verification plans that introduce a hierarchy of behavioral modeling and verification-oriented test benches target the problem in a conventional way by changing the methodology how existing tools are used [4]. While on block level formal methodologies could be introduced into industrial flows quickly with the support of EDA-vendors, increasing the complexity of the system under verification is continuously decreasing the applicability of formal methods, as presented in [2].

Another emerging verification approach attempts to formalize the property specification and evaluation of conventional simulation results introducing assertions. Derived from the digital domain, assertion-based verification automates the evaluation of simulation results and hence enables regression testing. A recent approach to include analog assertion-based verification on commercial platforms proposes property specification by implementing either an analog extension to System Verilog Assertions (SVA) or a library of analog assertion objects for the Open Verification Library (OVL) [19].

The tool AMT proposed in [20] uses a syntax extension called Signal Temporal Logic (STL) as an extension to the Property Specification Language (PSL) in order to verify properties on analog transient simulation waveforms. Due to a boolean layer created from threshold crossings of the analog signal, the expressiveness is mostly directed to timing verification and was demonstrated on a flash memory case-study. By formulating analog specifications in form of recurrence equations directly by incorporating PSL sequential extended regular expressions (SERE) as proposed in [27], again only very abstract, timing-oriented properties can be expressed.

Offering a specification methodology for complex analog properties, the application of ASL to transient simulation waveforms in [24] allows to specify and verify analog properties such as slew rate, oscillation and overshoot.

4 Unifying and Formalizing Analog Verification Methodologies

4.1 Discrete State Space Modeling

To apply the verification algorithms presented in the following sections, the analog circuit description has to be transferred into a discrete graph data structure that will be referred to as discrete analog transition structure (DATS).

Definition 1. (Discrete Analog Transition Structure (DATS))
For a set of atomic state propositions *AP* and a set of atomic transition propositions *TP*, the DATS M_{DATS} over AP, TP is a six tuple $M_{\text{DATS}} = (\Sigma, R, L_A, L_V, T, L_T)$ where

- Σ is a finite set of states of the system.
- $R \subseteq \Sigma \times \Sigma$ is a total transition relation, hence for every state $\sigma \in \Sigma$ there exists a state σ' such that $R(\sigma, \sigma')$.
- $L_A : \Sigma \rightarrow 2^{AP}$ is a labeling function that labels each state with the set of atomic propositions that are true in that state.
- $L_V : \Sigma \rightarrow \mathbb{R}^m$ is a labeling function that labels each state with the vector of m variables containing the values in this state of the extended state space variables and the algebraic variables of the differential algebraic equation (DAE) system.
- $T : R \rightarrow \mathbb{R}_0^+$ is a labeling function that labels each transition from σ to σ' with a real valued positive or zero transition time that represents the time required for the trajectory in the extended state space between these states.
- $L_T : R \rightarrow 2^{TP}$ is a labeling function that labels each transition with the set of atomic transition propositions that are true for that transition.

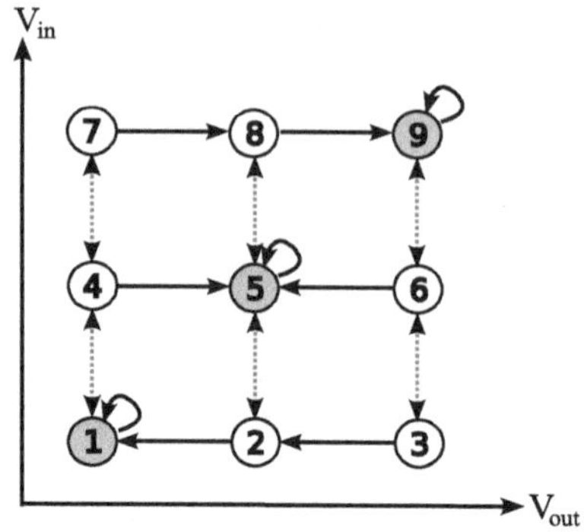

Fig. 5 Schematic illustration of a graph structure representing a DATS

Within the structure M_{DATS}, a path beginning at state σ is a sequence of states $\pi = \sigma_0\sigma_1\sigma_2\ldots\sigma_n$ with $\sigma_0 = \sigma$ and $R(\sigma_i, \sigma_{i+1})$ for $0 \leq i < n$.

Figure 5 illustrates a schematic graph structure representing a DATS with nine vertices modeling an imaginary analog circuit. The DC-operating-points of the modeled circuit are represented by vertices 1, 5 and 9. Thus, they have a loop transition to themselves stating that the circuit stays in this steady state infinitely until a change of input occurs. These transitions have zero transition time. A transition induced by an input change is modeled by a bidirectional edge, implying that this transition can only be taken if there is an input variable change in the corresponding extended state space dimension. Any non-steady state of the system has outgoing directed edges representing the dynamic behavior of the circuit.

For discrete modeling of an analog circuit as a DATS, a nonlinear first order DAE system

$$\mathbf{f}(\dot{\mathbf{x}}(t), \mathbf{x}(t), \mathbf{u}(t)) = \mathbf{0} \tag{1}$$

representing the input vector $\mathbf{u}(t)$ and the vector of the system variables $\mathbf{x}(t)$ is set up by applying modified nodal analysis (MNA) to the circuit netlist. The state space of the resulting DAE system, spanned by the system variables of the energy-storing elements $\mathbf{x}_e(t)$ and the input variables, is now divided into hypercubes of homogeneous behavior by comparing length and angle of the system variables' derivatives $\dot{\mathbf{x}}_e(t)$. The derivatives also define the transition relation of the hypercubes and the transition time. Finally, a graph data structure is generated from this information considering each hypercube as a vertex of the graph, connected according to the transition relation. The vertices are labeled with the state space variable values at the center of the corresponding hypercube and the transition times are the edge-labels of the DATS. Figure 6 illustrates the discrete modeling process.

The described modeling process is performed by the analog modeling and verification (AMV) tool [13, 14]. The verification approaches presented in the following are interfacing with this modeling tool and work on the generated discrete model.

4.2 *Verification of Analog System Properties Using the Analog Specification Language (ASL)*

In contemporary analog design flows, verification is not formalized and so is the specification. Test bench-based characterization of circuit properties is comparing an informal specification, often given in form of a table, of allowed ranges of certain circuit properties such as maximum power consumption, slew rate, startup time, etc. to experimental evaluation of the design under verification in a circuit test bench. Hence, there is no machine readable formalized specification methodology that guarantees a standardized verification approach. In fact, the incomplete and indirect specification of circuit properties leaves considerable freedom of interpretation to transfer the designer's intent into a circuit design as well as how to verify the specified properties.

While in modern test benches a set of simulation measurements is predefined in order to quickly characterize a given circuit, global quality management of the specification and verification is not yet widely introduced into design flows. Hence, without a formalized detailed specification, several sources for design errors exist.

There are two fundamental domains of property specification which have a high impact on how the verification of the properties can be performed. On the one hand, for today's design flows, a signal-oriented temporal property specification is needed in order to describe the desired behavior of simulation results. On the other hand, formal verification methodologies for analog circuits consider circuit behavior to be analyzed in the circuit's state space. Therefore, specification of dynamic temporal analog properties for application in formal verification flows is required to consider the state space domain.

Up to now, property specification approaches such as CTL-AT [9] for formal verification of analog circuits were based on extended versions of temporal logics. But due to its origin in temporal logics, CTL is not capable of offering a designer-oriented specification methodology. In order to gain acceptance for formal approaches in analog verification, a new methodology of property specification is necessary. While the Property Specification Language (PSL) [7] offered this step towards designer-oriented specification in the digital domain, the first approach to analog specification with PSL [20], as mentioned in Sect. 3.3, is only covering signal-based properties for assertion-based verification. Hence, it is not suited for describing properties of analog systems in state space.

4.2.1 ASL Language Concept

ASL syntax has been designed to be semantically deductive and therewith it shall reduce the time needed for understanding existing specifications in comparison to temporal logic specification [23, 26]. This is achieved by providing the possibility of creating parameterized macro functions involving a macro preprocessor. Thus, specification code can be sourced out to macro libraries allowing encapsulation and

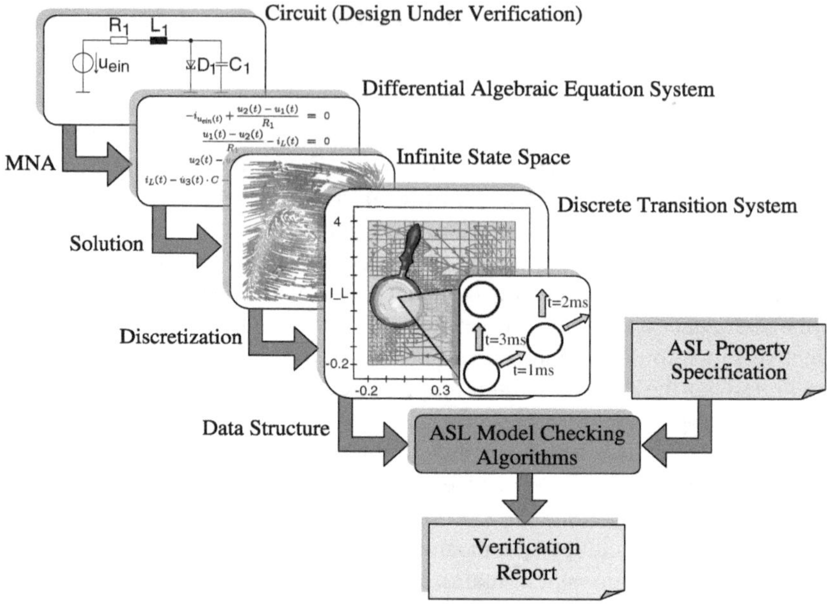

Fig. 6 ASL model checking flow with generation of a discrete model for nonlinear analog systems

reuse of specification code. Application of ASL specifications and algorithms is not restricted to state space models generated with the approach described in Sect. 4.1, so it can be adopted to other finite state machine-based modeling approaches like [8]. Moreover, as will be shown in Sect. 4.3, ASL can be used to specify and verify properties on conventional transient simulation waveforms by transferring them to a state space representation.

From the point of view of analog circuit developers, analog properties are represented by continuous physical values and their alteration over time. Thus, it is necessary to select states by calculations on their state space parameter values. Whether a state belongs to a set is decided by comparing the result of an arithmetic calculation to a specified interval. Extended path operations abstract from the reachability analysis concept of temporal logics and allow examination of more complex properties on paths within state space. As demonstrated in [23], the specification and therewith the verification possibilities of ASL are significantly extended compared to CTL-AT.

4.2.2 Model Checking with ASL

The goal of ASL model checking is to prove that properties specified in ASL are satisfied by the analog circuit modeled as DATS. Figure 6 illustrates the discrete modeling process and the application of ASL model checking algorithms. In order to evaluate the ASL specification on the DATS, an implementation of the verification algorithms identifies the sets of states and the properties of the state space variables on these sets and writes the verification results to a verification protocol. A first example of the ASL verification methodology was presented in Sect. 2. In the following, a specification template is demonstrated giving an example to identify overshooting behavior of a circuit. Firstly, the set `reachable` of reachable states has to be identified. The reachable states are those states that the circuit can adopt either from DC-operating-points or another stable operating point such as a periodic steady state for a given supply voltage. The set `DCpoints` is precomputed and contains the states of the DATS that represent DC-operating-points:

```
reachable = reach from DCpoints;
```

In the next step, the set of states that violates the specification has to be identified. For overshooting of the output variable `V_out` of a circuit, a number variable `%max_V_out` has to be assigned with a maximum acceptable value. Every state of the circuit where its state space variable `V_out` is above `%max_V_out` is a state that violates the specification:

```
overshoot_states = on reachable select value(V_out)[> %max_V_out];
```

The specification is satisfied if the set `overshoot_states` is empty. Hence, the following assertion can be formulated in ASL.

```
for is_empty(overshoot_states) assert true;
```

If the assertion is false, a counterexample represented by an analog signal trace in form of a piecewise linear input stimulus for the input variable `V_q` can be generated, as will be described in Sect. 4.4:

```
counterexample(V_q) from DCpoints to overshoot_states;
```

4.3 Applying ASL Verification Algorithms to Transient Simulation Waveforms

In order to obtain a wider field of application for ASL property specification and evaluation and to develop another formal property verification methodology in Sect. 4.5, evaluation of ASL property specifications shall be extended from discrete state space models to analog transient simulation waveforms. Therefore, the simulation waveforms have to be transferred into a state space representation to be introduced into the ASL verification tool-chain.

Transient simulation data consists of data tuples containing a sequence of signal values and their corresponding time points. If each tuple is considered as a vertex of

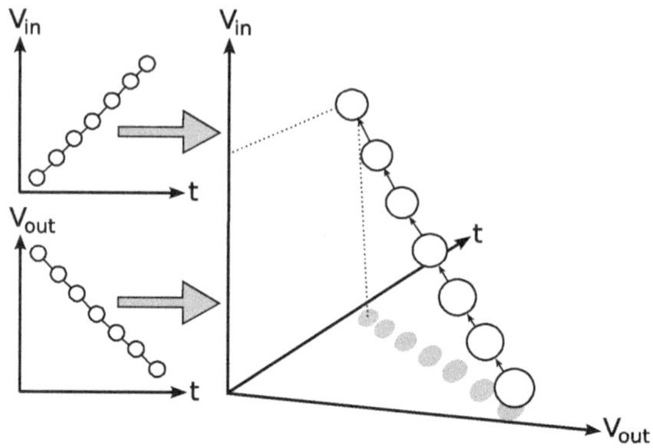

Fig. 7 Graph structure obtained from transferring transient simulation waveforms into a DATS model

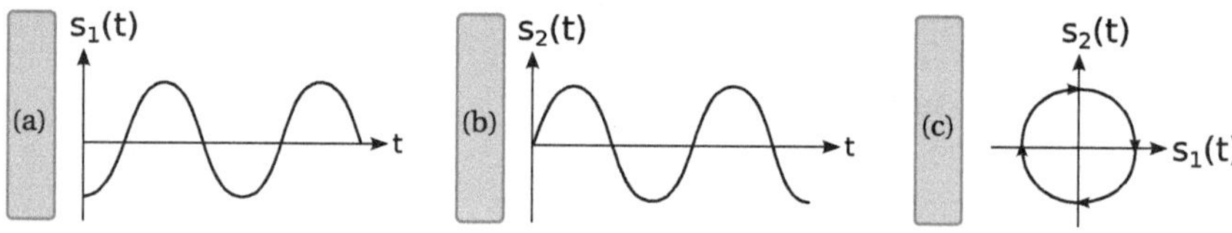

Fig. 8 Periodic transient signal waveforms $s_1(t)$ (**a**) and $s_2(t)$ (**b**). State space representation of periodic signals $s_1(t)$ and $s_2(t)$ (**c**)

a graph structure with adjacent tuples connected by a directed transition edge and marked with the difference of the time points as transition time, a representation of the signals in state space can be created as depicted in the schematic illustration in Fig. 7. Due to the transition times now being defined by the labels of the graph transitions, a time axis is obsolete but plotted for better understanding. As illustrated in Figs. 8a and 8b for two periodic signals, their combined state space representation in Fig. 8c forms a circle by mapping both signals to a plot over axis $s_1(t)$ and $s_2(t)$. For the state space representation of transient signals, a detection of periodic behavior is necessary in order to create closed cycles for oscillation detection. With a defined tolerance interval, for each vertex is checked whether its coordinate vector maps to a preceding one and transitions are connected accordingly.

4.4 Counterexample Generation

The specification of analog circuit properties using the ASL methodology introduced in Sect. 4.2 allows to identify regions in the state space of an analog circuit that violate the specification. In order to understand how the circuit behavior reaches such a region, a counterexample can be generated.

Definition 2. (Counterexample on a DATS)
A counterexample for the analog circuit model represented as DATS is a path π_{ce} from a defined starting state σ_0 to the set of states Φ violating the specification:

$$\pi_{ce} = \sigma_0 \ldots \sigma_i : \exists\, i \geq 0 : \sigma_i \in \Phi \tag{2}$$

Due to the structure of the DATS, on such paths, for every extended state space variable the values of this variable and the corresponding transition time can be recorded for every state transition on the path. This yields a piecewise linear signal trace over time which can then be visually inspected by the verification engineer. Moreover, by generating such a signal trace for every input variable of the circuit, piecewise linear input stimuli are obtained which can be directly applied to a circuit test bench in order to inspect the formal verification results within the conventional simulation environment.

4.5 Complete State Space-Covering Input Stimuli Generation

Transient simulation results are highly depending on the circuit designer to find the right input stimuli to identify critical system behavior. Obviously, model or equivalence checking can resolve this problem. However, transient simulation is the state of the art in industrial design flows and introduction of completely new methodologies will take some time. Therefore, profiting of formal techniques by generation of input stimuli for transient simulation that cover the complete state space of an analog circuit is possible without fundamentally changing the design flow.

Derived from the counterexample generation approach introduced in the previous section, input stimuli covering the complete reachable area of the state space of the analog circuit can be computed by traversing the graph modeling an analog circuit as a DATS. Basically, the idea is to visit every reachable state of the graph structure and recording the input values and accumulated times of traveled edges during graph traversal. This concept corresponds to generating a piecewise linear counterexample input stimulus for every reachable state of the DATS model [22, 25].

Definition 3. (Complete State Space-Covering Input Stimuli on a DATS)
A complete state space-covering input stimulus for the analog circuit model represented as a DATS is a path π_{is} from a defined starting state σ_0 that visits every state of the set of reachable states Φ of the DATS. The set Π_{is} consists of all π_{is} satisfying this requirement.

$$\Pi_{is} = \left\{ \pi_{is} \middle| \exists n : \pi_{is} = \sigma_0 \ldots \sigma_n \wedge \bigcup_{0 \leq i \leq n} \sigma_i = \Phi \right\} \tag{3}$$

$$\pi_{is} \in \Pi_{is} \tag{4}$$

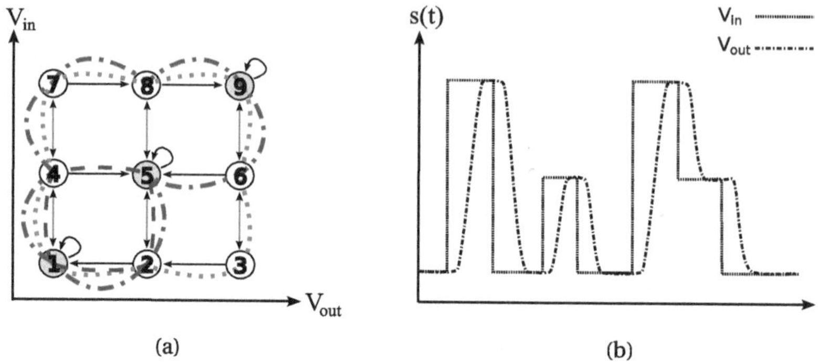

Fig. 9 Path generated by the stimuli generation algorithm (**a**) and the corresponding input/output behavior (**b**)

As illustrated in Fig. 9a, the algorithm starts at a vertex representing a DC-operating-point of the circuit and then follows the edges of the graph structure from vertex to vertex, storing the corresponding input signal values at the visited vertices and the transition times. Therewith, tuples of value and time are obtained, representing a piecewise linear input stimulus for each input of the system. These input stimuli can now be used in a circuit simulator, bringing the circuit to every reachable state during simulation, resulting in complete state space coverage of the simulation as illustrated in Fig. 9b. Moreover, a formal assertion-based verification can be performed by transient simulation of a circuit using complete state space-covering input stimuli and evaluating the results with the ASL property verification methodology on the resulting simulation waveforms as introduced in Sect. 4.3.

4.6 Equivalence Checking Methodology Using Complete State Space-Covering Input Stimuli

The simulation of a single circuit using complete state space-covering input stimuli can reveal corner case behavior not identified by user-defined input stimuli. Furthermore, an equivalence checking methodology for analog circuits based on the stimuli generation approach can be developed, giving certainty about the level of equality between two circuit implementations [25]. The idea is to generate a stimulus for the system that covers the system's complete state space during a transient simulation. If another circuit is simulated using the same input stimulus, the level of equivalence of the two systems is determined by the level of deviation of the transient responses of the two circuits.

For each of the two circuits to compare, in the following referred to as circuit A and circuit B, complete state space-covering input stimuli are generated for every input of the circuit. Subsequently, four simulation runs are needed. Circuit A is

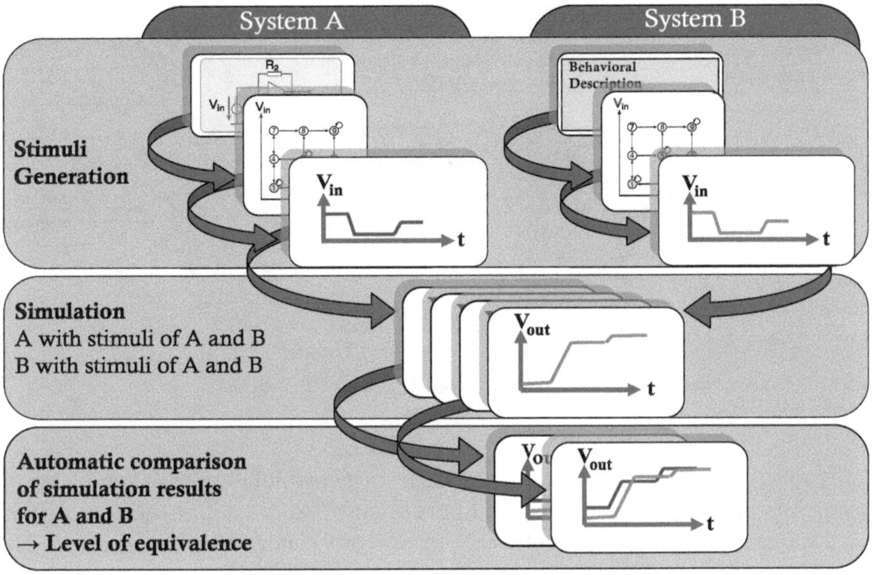

Fig. 10 Equivalence checking flow using complete state space-covering input stimuli

simulated with stimuli of A and B, followed by simulating circuit B with stimuli of A and B. The simulation results are automatically compared using an error measure, reporting equivalence if a user-specifiable maximum error value is not exceeded.

If circuits A and B show equivalent behavior when simulated with stimuli generated from circuit A, then the complete behavior of circuit A is included in circuit B: $A \subseteq B$. If circuit A and B show equivalent behavior for simulation with stimuli generated from circuit B, then the complete behavior of circuit B is included in circuit A: $B \subseteq A$. If both conditions hold, circuit A and circuit B are considered as equivalent with respect to the user-defined maximum error:

$$A \subseteq B \ \wedge \ B \subseteq A \implies A \approx B \tag{5}$$

In practical applications, often only one direction of the proof may be necessary. Figure 10 illustrates the described equivalence checking methodology.

4.7 The Verification Flow Perspective

Around the specification and verification possibilities based on ASL and the complete state space-covering stimuli generation algorithm, an unified framework of analog formal verification methodologies has been developed, which will be presented in the following. While fundamental formal methods such as model checking and equivalence checking are enhanced by the ASL verification algorithms, com-

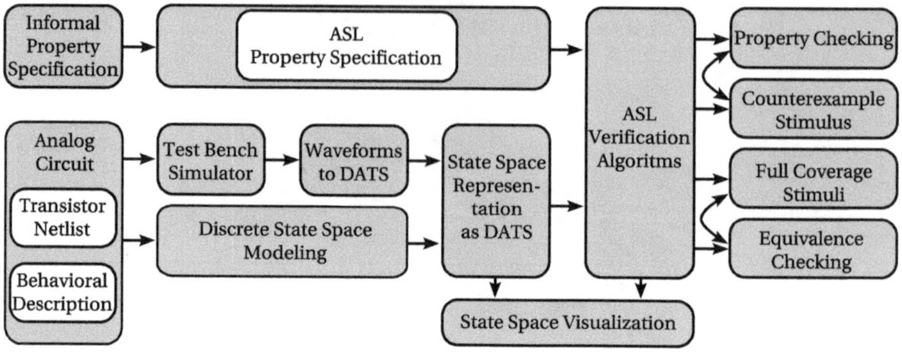

Fig. 11 ASL verification flow for different verification methodologies

bining new verification approaches such as transient simulation with complete state space-covering input stimuli and the application of ASL specifications to conventional transient simulation waveforms offers a set of new verification methodologies.

Figure 11 illustrates the possibilities of the ASL-based verification framework offering several combinations of different modeling and verification approaches. Starting from a behavioral or transistor netlist representation of an analog circuit, a discrete state space model for each circuit under verification is generated. Alternatively, conventional transient simulation results from test bench-based verification can be transferred into a partial state space representation, which then can be processed by the verification algorithms accordingly. The following verification methodologies can then be applied.

- The state space model can be checked against an ASL property specification which is the method of property checking or model checking.
- From identified state space regions that violate the specification, counterexample stimuli can be generated, which then in turn can be used in a test bench simulation environment to analyze the specification-violating behavior.
- By a systematic traversal of the state space model, complete state space covering piecewise linear input stimuli can be generated. Based on these complete-coverage input stimuli, the three following methodologies are available.
  - The complete-coverage stimuli can be used in test bench-based simulation environments for manual analysis of the transient response to these stimuli that guide the simulator into every reachable state of the analog circuit.
  - The transient response to such complete-coverage stimuli can be re-transferred into a state space representation for evaluation of assertions specified in ASL. This is an alternative complete, and therewith formal, property verification approach with the advantage of directly operating on the results of the transient simulation which improves accuracy of measured values as the error of the discrete modeling process is not existent.

– By generating complete-coverage stimuli for two circuits under verification and then comparing the deviation of the transient responses to the stimuli of both circuits using a specific ASL specification, a formal equivalence checking methodology is given due to the transient responses of both circuits being directed to every reachable state.

5 Experimental Results

With the verification of the oscillator circuit presented in Sect. 2, first results of the ASL property verification methodology were demonstrated. For identification of the bad initial conditions from which the circuit does not begin oscillating, the ASL specification shown in Listing 1 was used.

Listing 1 ASL specification for oscillator verification

```
# Assert that circuit has no non−periodic steady states
for is_empty(DCpoints) assert true;

# Which states lead into DCpoints?
bad_initial_conditions = reach DCpoints;
```

For this example, the state space model generation with 1970 states takes 54 seconds. The ASL model checking algorithms have a runtime of 3 seconds. All examples were computed on a single core of a Pentium 4 D with a clock frequency of 2.8 GHz and 2 GB of RAM.

In order to present the combination of property specification and counterexample generation, the overshoot specification and verification methodology presented in Sect. 4.2.2 was applied to a Sallen-Key biquad lowpass filter with a cut-off frequency of 1000 Hz. The circuit schematic is illustrated in Fig. 12a with $C_1 = 5$ nF, $C_2 = 50$ nF, $R_1 = R_2 = R_4 = 10\,\mathrm{k}\Omega$, $R_3 = 500\,\Omega$. The input voltage range is ± 1 V and the operational amplifier has a supply voltage of ± 2 V. Figure 12b shows the counterexample input stimulus V_{in} generated by the ASL verification algorithms with the overshooting behavior to -1.47 V of the transient response of V_{out} identified by the ASL expression:

```
counterexample(V_in) from DCpoints to reachable and V_out[< -1.4]
```

Model generation with 987 reachable states took 19 seconds and the verification and counterexample generation took 1.2 seconds.

As an alternative to the model checking methodology with counterexample generation, a complete state space-covering input stimulus as introduced in Sect. 4.5 was generated. The transient simulation response was then transferred back into a

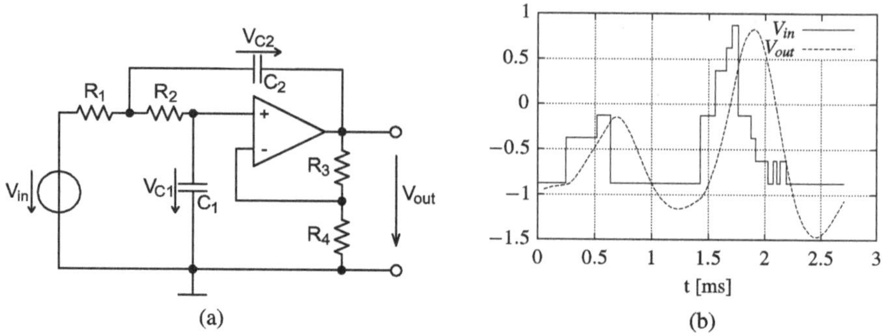

Fig. 12 Circuit schematic of the Sallen-Key biquad lowpass filter (**a**). Counterexample input stimulus and transient response leading to overshooting of the output to -1.47 V

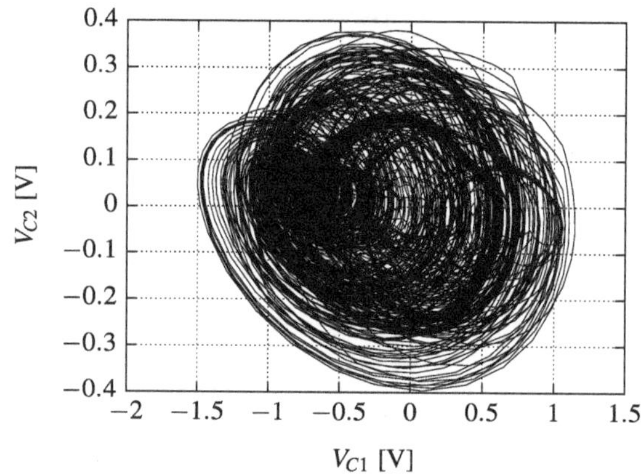

Fig. 13 Transient response to the generated input stimulus of the Sallen-Key biquad lowpass filter plotted over V_{C1} and V_{C2} proving complete coverage of the reachable area

state space representation as described in Sect. 4.3 for application of the ASL overshoot property specification. Therewith, another automated verification methodology is given that detects the overshooting behavior of the Sallen-Key circuit. Figure 13 shows the transient response to the complete state space-covering input stimulus plotted over V_{C1} and V_{C2}, densely covering all the reachable value combinations of V_{C1} and V_{C2}. The stimulus generation with an overall time length of 758 ms and 9191 time steps took 11 seconds on the discrete model.

In addition, the complete state space-covering input stimulus was used to equivalence-check the presented circuit against a behavioral model where no overshooting behavior was implemented. The comparison of the transient responses of both circuits to the input stimulus revealed the differences where the two implementations have non-equal behavior.

6 Conclusions

In this chapter, an insight into new methods for formal verification of analog circuits was provided. The success of formal methods in the digital domain is inevitable. Thus, in order to keep pace, analog and especially mixed-signal circuit design flows require solutions for formalizing the verification effort.

Although there are several challenges to solve until an industrial application can be considered, the presented verification methodologies around the Analog Specification Language (ASL) and the corresponding verification algorithms show that formal verification for analog circuits is possible. While the application of ASL to transient simulation waveforms could be a first step towards formalizing analog verification, the methodologies of complete state space-covering input stimuli generation for property and equivalence checking, model checking, and counterexample generation offer a variety of formal solutions for different applications.

References

1. G. Al-Sammane, M. H. Zaki, and S. Tahar. A symbolic methodology for the verification of analog and mixed signal designs. In *DATE '07: Proceedings of the Conference on Design, Automation and Test in Europe*, pages 249–254, EDA Consortium, San Jose, CA, USA, 2007.
2. E. Barke, D. Grabowski, H. Graeb, L. Hedrich, S. Heinen, R. Popp, S. Steinhorst, and Y. Wang. Formal approaches to analog circuit verification. In *DATE '09: Proceedings of the Conference on Design, Automation and Test in Europe*, pages 724–729, 2009.
3. E. M. Clarke and E. A. Emerson. Design and synthesis of synchronization skeletons using branching-time temporal logic. In *Logic of Programs*, volume 131 of *Lecture Notes in Computer Science*, pages 52–71. Springer-Verlag, London, 1982.
4. H. Chang and K. Kundert. Verification of complex analog and RF IC designs. *Proceedings of the IEEE*, 95(3):622–639, 2007.
5. T. Dastidar and P. Chakrabarti. A verification system for transient response of analog circuits using model checking. In *VLSID '05: Proceedings of the 18th International Conference on VLSI Design held jointly with 4th International Conference on Embedded Systems Design*, pages 195–200, 2005.
6. T. R. Dastidar and P. P. Chakrabarti. A verification system for transient response of analog circuits. *ACM Trans. Des. Autom. Electron. Syst.*, 12(3):1–39, 2007.
7. H. Foster, E. Marschner, and Y. Wolfsthal. IEEE 1850 PSL: The Next Generation. 2005. URL: `http://www.pslsugar.org/papers/ieee1850psl-the_next_generation.pdf`.
8. M. Freibothe, J. Schönherr, and B. Straube. Formal verification of the quasi-static behavior of mixed-signal circuits by property checking. *Electr. Notes Theor. Comput. Sci.*, 153(3):23–35, 2006.
9. D. Grabowski, D. Platte, L. Hedrich, and E. Barke. Time constrained verification of analog circuits using model-checking algorithms. In *FAC 2005: Proceedings of the First Workshop on Formal Verification of Analog Circuits*, pages 37–52, 2005.
10. A. Ghosh and R. Vemuri. Formal verification of synthesized analog designs. In *ICCD '99: Proceedings of the 1999 IEEE International Conference on Computer Design*, page 40, IEEE Computer Society, Washington, DC, USA, 1999.

11. L. Hedrich and E. Barke. A formal approach to nonlinear analog circuit verification. In *ICCAD '95: Proceedings of the 1995 IEEE/ACM International Conference on Computer-Aided Design*, pages 123–127, IEEE Computer Society, Washington, DC, USA, 1995.
12. L. Hedrich and E. Barke. A formal approach to verification of linear analog circuits with parameter tolerances. In *DATE '98: Proceedings of the Conference on Design, Automation and Test in Europe*, pages 649–655, IEEE Computer Society, Washington, DC, USA, 1998.
13. W. Hartong, L. Hedrich, and E. Barke. Model checking algorithms for analog verification. In *Proceedings of the 39th Conference on Design Automation (DAC '02)*, pages 542–547, 2002.
14. W. Hartong, R. Klausen, and L. Hedrich. Formal verification for nonlinear analog systems: Approaches to model and equivalence checking. 2004.
15. K. D. Jones, J. Kim, and V. Konrad. Some 'real world' problems in the analog and mixed signal domains. In Gordon J. Pace and Satnam Singh, editors, *Seventh International Workshop on Designing Correct Circuits: Budapest, 29–30 March 2008: Participants' Proceedings*, pages 15–29. ETAPS 2008, March 2008. A Satellite Event of the ETAPS 2008 group of conferences.
16. R. P. Kurshan and K. L. McMillan. Analysis of digital circuits through symbolic reduction. *IEEE Trans. on CAD of Integrated Circuits and Systems*, 10(11):1356–1371, 1991.
17. S. Little, N. Seegmiller, D. Walter, C. Myers, and T. Yoneda. Verification of analog/mixed-signal circuits using labeled hybrid petri nets. In *ICCAD '06: Proceedings of the 2006 IEEE/ACM International Conference on Computer-Aided Design*, pages 275–282, ACM, New York, NY, USA, 2006.
18. C. J. Myers, R. R. Harrison, D. Walter, N. Seegmiller, and S. Little. The case for analog circuit verification. *Electr. Notes Theor. Comput. Sci.*, 153(3):53–63, 2006.
19. R. Mukhopadhyay, S. K. Panda, P. Dasgupta, and J. Gough. Instrumenting ams assertion verification on commercial platforms. *ACM Trans. Des. Autom. Electron. Syst.*, 14(2):1–47, 2009.
20. D. Nickovic and O. Maler. Amt: A property-based monitoring tool for analog systems. In J.-F. Raskin and P. S. Thiagarajan, editors, *FORMATS*, volume 4763 of *Lecture Notes in Computer Science*, pages 304–319. Springer, 2007.
21. S. Seshadri and J. A. Abraham. Frequency response verification of analog circuits using global optimization techniques. *J. Electron. Test.*, 17(5):395–408, 2001.
22. S. Steinhorst and L. Hedrich. A formal approach to complete state space-covering input stimuli generation for verification of analog systems. In *Analog 2008: 10. ITG/GMM-Fachtagung Entwicklung von Analogschaltungen mit CAE-Methoden*, 2008.
23. S. Steinhorst and L. Hedrich. Model checking of analog systems using an analog specification language. In *Proc. of the Conference on Design, Automation and Test in Europe 2008 (DATE '08)*, pages 3247–329, 2008.
24. S. Steinhorst and L. Hedrich. Joint property specification for transient simulation and formal verification of analog circuits. In *Proc. of the edaWorkshop'09*, pages 13–18, VDE Verlag, Berlin, 2009.
25. S. Steinhorst and L. Hedrich. Improving verification coverage of analog circuit blocks by state space-guided transient simulation. In *Circuits and Systems, 2010. ISCAS 2010. IEEE International Symposium on*, May 2010.
26. S. Steinhorst, A. Jesser, and L. Hedrich. Advanced property specification for model checking of analog systems. In *Analog 2006: 9. ITG/GMM-Fachtagung Entwicklung von Analogschaltungen mit CAE-Methoden*, pages 63–68, 2006.
27. G. Al Sammane, M. H. Zaki, Z. J. Dong, and S. Tahar. Towards assertion based verification of analog and mixed signal designs using PSL. In *FDL*, pages 293–298. ECSI, 2007.

Index

P. Li et al. (eds.), *Simulation and Verification of Electronic and Biological Systems*,
DOI 10.1007/978-94-007-0149-6, © Springer Science+Business Media B.V. 2011

FSC
www.fsc.org
MIX
Papier | Fördert
gute Waldnutzung
FSC® C083411

Kolibri 360 GmbH
Ferdinand-Jühlke-Straße 7
99095 Erfurt, Deutschland
produktsicherheit@kolibri360.de